T0137997

Solutions Manual for Lang's

Linear Algebra

Springer
New York
Berlin
Heidelberg
Barcelona
Budapest
Hong Kong
London
Milan
Paris
Santa Clara
Singapore
Tokyo

Rami Shakarchi

Solutions Manual for Lang's
Linear Algebra

Rami Shakarchi
P.O. Box 205612 Yale Station
New Haven, CT 06520
USA

Mathematics Subject Classification (1991): 15-01

Printed on acid-free paper.

Production managed by Lesley Poliner; manufacturing supervised by Joe Quatela.
Camera-ready copy prepared by the author.
Printed and bound by Braun-Brumfield, Inc., Ann Arbor, MI.
Printed in the United States of America.

9 8 7 6 5 4 3 2 1

ISBN 0-387-94760-4 Springer-Verlag New York Berlin Heidelberg SPIN 10536794

To my parents

Preface

The present volume contains all the exercises and their solutions of Lang's *Linear Algebra.* Solving problems being an essential part of the learning process, my goal is to provide those learning and teaching linear algebra with a large number of worked out exercises. Lang's textbook covers all the topics in linear algebra that are usually taught at the undergraduate level: vector spaces, matrices and linear maps including eigenvectors and eigenvalues, determinants, diagonalization of symmetric and hermitian maps, unitary maps and matrices, triangulation, Jordan canonical form, and convex sets. Therefore this solutions manual can be helpful to anyone learning or teaching linear algebra at the college level.

As the understanding of the first chapters is essential to the comprehension of the later, more involved chapters, I encourage the reader to work through all of the problems of Chapters I, II, III and IV. Often earlier exercises are useful in solving later problems. (For example, Exercise 35, §3 of Chapter II shows that a strictly upper triangular matrix is nilpotent and this result is then used in Exercise 7, §1 of Chapter X.) To make the solutions concise, I have included only the necessary arguments; the reader may have to fill in the details to get complete proofs.

Finally, I thank Serge Lang for giving me the opportunity to work on this solutions manual, and I also thank my brother Karim and Steve Miller for their helpful comments and their support.

Rami Shakarchi
Yale, 1996

I thank Rami Shakarchi very much for having prepared this answer book.

Serge Lang
Yale, 1996

Contents

CHAPTER I

Vector Spaces

I, §1 Definitions

1. *Let V be a vector space. Using the properties* **VS 1** *through* **VS 8**, *show that if c is a number then $cO = O$.*

SOLUTION. We have $cO = c(O+O) = cO + cO$, but we also have $cO = O + cO$, hence

$$cO + cO = O + cO.$$

Adding $(-cO)$ to both sides shows that $cO = O$.

2. *Let c be a number $\neq 0$, and v an element of V. Prove that if $cv = O$, then $v = O$.*

SOLUTION. Exercise 1 implies, $O = (1/c)O = (1/c)(cv) = (c/c)v = 1v = v$.

3. *In the vector space of functions, what is the function satisfying the condition* **VS 2**?

SOLUTION. The zero function, namely, $f(x) = 0$ for all x plays the role of the identity.

4. *Let V be a vector space and v, w two elements of V. If $v + w = O$, show that $w = -v$.*

SOLUTION. We have $-w = -w + O = -w + (v + w) = v + w - w = v$.

5. *Let V be a vector space, and v, w be two elements of V such that $v + w = v$. Show that $w = 0$.*

SOLUTION. We know that $v + O = v$ so $v + O = v + w$. Adding $-v$ to both sides shows that $O = w$.

6. *Let* A_1, A_2 *be vectors in* $\mathbf{R}^n$. *Show that the set of all vectors B in* $\mathbf{R}^n$ *such that* ***B*** *is perpendicular to both* A_1 *and* A_2 *is a subspace.*

SOLUTION. See Exercise 7.

7. *Generalize Exercise 6, and prove: Let* $A_1,\ldots, A_r$ *be vectors in* $\mathbf{R}^n$. *Let W be the set of vectors B in* $\mathbf{R}^n$ *such that* $B \cdot A_i = 0$ *for every* $i = 1,\ldots, r$. *Show that W is a subspace of V.*

SOLUTION. The definition of the dot product implies $O \cdot A_i = 0$ for all i; thus $O \in W$. If B_1 and B_2 lie in W, then the properties of the inner product imply

$$A_i \cdot (B_1 + B_2) = A_i \cdot B_1 + A_i \cdot B_2 = 0 + 0 = 0 \quad \text{and} \quad A_i \cdot (cB_1) = c(A_i \cdot B_1) = 0$$

for all i; thus W is a subspace of $\mathbf{R}^n$.

8. *Show that the following sets of elements in* $\mathbf{R}^2$ *form subspaces.*
(a) The set of all (x, y) *such that* $x = y$.
(b) The set of all (x, y) *such that* $x - y = 0$.
(c) The set of all (x, y) *such that* $x + 4y = 0$.

SOLUTION. In each case let W be the set in question.
(a) Since $0 = 0$, we have $O \in W$. Clearly, $x_1 + x_2 = y_1 + y_2$ and $cx_1 = cy_1$ whenever (x_1, y_1), $(x_2, y_2) \in W$ and $c \in \mathbf{R}$.

(b) We have $0 - 0 = 0$, so $O \in W$. If (x_1, y_1), $(x_2, y_2) \in W$ and $c \in \mathbf{R}$, then

$$(x_1 + x_2) - (y_1 + y_2) = (x_1 - y_1) + (x_2 - y_2) = 0$$

and

$$cx_1 - cy_1 = c(x_1 - y_1) = 0,$$

so W is a subspace.

(c) Since $0 + 4 \times 0 = 0$, we have $O \in W$. If (x_1, y_1), $(x_2, y_2) \in W$ and $c \in \mathbf{R}$, then

$$(x_1 + x_2) + 4(y_1 + y_2) = (x_1 + 4y_1) + (x_2 + 4y_2) = 0 \quad \text{and} \quad cx_1 + 4cy_1 = 0,$$

so W is a subspace.

9. *Show that the following sets of elements of* $\mathbf{R}^3$ *form subspaces.*

(a) The set of all (x, y, z) *such that* $x + y + z = 0$.

(b) The set of all (x, y, z) *such that* $x = y$ *and* $2y = z$.

(c) The set of all (x, y, z) *such that* $x + y = 3z$.

SOLUTION. In each case let W be the set in question.

(a) We have $O \in W$ because $0+0+0=0$. If $(x_1, y_1, z_1), (x_2, y_2, z_2) \in W$, then

$$x_1 + x_2 + y_1 + y_2 + z_1 + z_2 = 0 \quad \text{and} \quad cx_1 + cy_1 + cz_1 = 0$$

so W is a subspace.

(b) We have $0 = 0$ and $2 \times 0 = 0$, so $O \in W$. If $(x_1, y_1, z_1), (x_2, y_2, z_2) \in W$ and c is a real number, then

$$x_1 + x_2 = y_1 + y_2, \quad 2(y_1 + y_2) = z_1 + z_2, \quad cx_1 = cy_1, \quad 2cy_1 = cz_1,$$

so W is a subspace.

(c) Clearly, $0+0=3\times 0$, so $O \in W$. If $(x_1, y_1, z_1), (x_2, y_2, z_2) \in W$ and $c \in \mathbf{R}$, then

$$x_1 + x_2 + y_1 + y_2 = 3(z_1 + z_2) \quad \text{and} \quad cx_1 + cy_1 = 3cz_1,$$

so W is a subspace.

10. *If U, W are subspaces of a vector space V, show that* $U \cap W$ *and* $U + W$ *are subspaces.*

SOLUTION. (i) Since $O \in U$ and $O \in W$, we have $O \in U \cap W$. If $v_1, v_2 \in U \cap W$, then $v_1, v_2 \in U$ so $v_1 + v_2 \in U$ and $v_1, v_2 \in W$, hence $v_1 + v_2 \in W$. Thus $v_1 + v_2$ belongs to $U \cap W$. Similarly, cv_1 belongs to U and W, so $cv_1 \in U \cap W$. Consequently, $U \cap W$ is a subspace of V.
(ii) Since O belongs to U and W, O belongs to $U + W$. If $a, b \in U + W$, then we can write $a = u_1 + w_1$ and $b = u_2 + w_2$, where $u_i \in U$ and $w_i \in W$. Since U and W are subspaces, we see that $a + b = (u_1 + u_2) + (w_1 + w_2)$ and $ca = cu_1 + cw_1$ belong to $U + W$, so $U + W$ is a subspace of V.

11. *Let K be a subfield of a field L. Show that L is a vector space over K. In particular,* **C** *and* **R** *are vector spaces over* **Q**.

SOLUTION .If x and y belong to L, then since L is a field, $x+y$ is an element of L. Moreover, if $c \in K$, then $c \in L$ so that $cx \in L$. The element O of **VS 2** is simply 0, and in **VS 3** we have $-x=(-1)x$. All the other axioms are verified at once.

12. *Let K be the set of all numbers which can be written in the form $a+b\sqrt{2}$ where a, b are rational numbers. Show that K is a field.*

SOLUTION. Clearly, K is a subset of the complex numbers. If $a+b\sqrt{2}$ and $c+d\sqrt{2}$ belong to K, then

$$\left(a+b\sqrt{2}\right)+\left(c+d\sqrt{2}\right)=(a+b)+(c+d)\sqrt{2}$$

and

$$\left(a+b\sqrt{2}\right)\cdot\left(c+d\sqrt{2}\right)=(ac+2bd)+(ad+bc)\sqrt{2}.$$

Since $\mathbf{Q}$ is a field, we see at once that K is closed under addition and multiplication. For the other properties, note that $-a-b\sqrt{2} \in K$ and that if $a+b\sqrt{2}=0$, then $a=b=0$; so if $a+b\sqrt{2} \neq 0$, we have

$$\left(a+b\sqrt{2}\right)^{-1}=\frac{1}{a+b\sqrt{2}}=\frac{a}{a^2-2b^2}-\frac{b}{a^2-2b^2}\sqrt{2}$$

which belongs to K. We simply multiplied the numerator and denominator by $a-b\sqrt{2}$. Finally, $0=0+0\sqrt{2} \in K$ and $1=1+0\sqrt{2} \in K$.

13. *Let K be the set of all numbers which can be written in the form $a+bi$ where a, b are rational numbers. Show that K is a field.*

SOLUTION. We see that if $a+bi$ and $c+di$ belong to K, then

$$(a+bi)+(c+di)=(a+b)+(c+d)i$$

and

$$(a+bi)\cdot(c+di)=(ac-bd)+(ad+bc)i.$$

But $\mathbf{Q}$ is a field, so K is closed under addition and multiplication. Moreover, $-a-bi \in K$, and if $a+bi=0$, then $a=b=0$; so if $a+bi \neq 0$, we get

$$(a+bi)^{-1} = \frac{1}{a+bi} = \frac{a}{a^2+b^2} - \frac{b}{a^2+b^2}i$$

which belongs to K. We multiplied the numerator and denominator by the conjugate of $a+bi$. Finally, we have $0=0+0i$ and $1=1+0i$, so 0 and 1 belong to K.

14. *Let c be a rational number >0, and let γ be a real number such that $\gamma^2 = c$. Show that the set of all numbers which can be written in the form $a+b\gamma$ where a, b are rational numbers, is a field.*

SOLUTION. Let K be the set of numbers we are considering. If γ is rational, then $K=\mathbf{Q}$. Suppose that γ is irrational and let $a+b\gamma \in K$ and $u+t\gamma \in K$, then

$$(a+b\gamma)+(u+t\gamma) = (a+u)+(b+t)\gamma$$

and

$$(a+b\gamma)\cdot(u+t\gamma) = (au+btc)+(at+bu)\gamma.$$

Since c is rational and $\mathbf{Q}$ is a field, we see that K is closed under addition and multiplication. Clearly, $-a-b\gamma \in K$. Suppose that $a+b\gamma \neq 0$. Since γ is irrational, $a-b\gamma \neq 0$, so we can divide by $a-b\gamma$, hence

$$(a+b\gamma)^{-1} = \frac{a-b\gamma}{(a+b\gamma)(a-b\gamma)} = \frac{a}{a^2-cb^2} - \frac{b}{a^2-cb^2}\gamma.$$

Finally, 0 and 1 belong to K because $0=0+0\gamma$ and $1=1+0\gamma$.

I, §2 Bases

1. *Show that the following vectors are linearly independent (over $\mathbf{C}$ or $\mathbf{R}$).*

(a) $(1,1,1)$ *and* $(0,1,-2)$ *(b)* $(1,0)$ *and* $(1,1)$

(c) $(-1,1,0)$ *and* $(0,1,2)$ *(d)* $(2,-1)$ *and* $(1,0)$

(e) $(\pi,0)$ *and* $(0,1)$ *(f)* $(1,2)$ *and* $(1,3)$

(g) $(1,1,0)$, $(1,1,1)$, *and* $(0,1,-1)$ *(h)* $(0,1,1)$, $(0,2,1)$, *and* $(1,5,3)$

SOLUTION. (a) If a and b are numbers such that $a(1,1,1)+b(0,1,-2)=O$, then we have

$$\begin{cases} a = 0 \\ a + b = 0 \\ a - 2b = 0 \end{cases}$$

so $a = b = 0$.

(b) If $a(1,0) + b(1,1) = O$, then

$$\begin{cases} a + b = 0 \\ b = 0 \end{cases}$$

so $a = b = 0$.

(c) If $a(-1,1,0) + b(0,1,2) = O$, then

$$\begin{cases} -a = 0 \\ a + b = 0 \\ 2b = 0 \end{cases}$$

so $a = b = 0$.

(d) If $a(2,-1) + b(1,0) = O$, then

$$\begin{cases} 2a + b = 0 \\ -a = 0 \end{cases}$$

so $a = b = 0$.

(e) If $a(\pi, 0) + b(0,1) = O$, then

$$\begin{cases} a\pi = 0 \\ b = 0 \end{cases}$$

so $a = b = 0$.

(f) If $a(1,2) + b(1,3) = O$, then

$$\begin{cases} a + b = 0 \\ 2a + 3b = 0. \end{cases}$$

The second equation minus twice the first implies $b = 0$. So $a = b = 0$.

(g) If $a(1,1,0)+b(1,1,1)+c(0,1,-1)=O$, then

$$\begin{cases} a+b=0 \\ a+b+c=0 \\ b-c=0 \end{cases}$$

The second equation minus the first implies $c=0$. So $a=b=c=0$.

(h) If $a(0,1,1)+b(0,2,1)+c(1,5,3)=O$, then

$$\begin{cases} c=0 \\ a+2b+5c=0 \\ a+b+3c=0 \end{cases}$$

Subtracting the third equation from the second, we see that $a=b=c=0$.

2. *Express the given vector X as a linear combination of the given vectors A, B and find the coordinates of X with respect to A, B.*
(a) $X=(1,0)$, $A=(1,1)$, $B=(0,1)$
(b) $X=(2,1)$, $A=(1,-1)$, $B=(1,1)$
(c) $X=(1,1)$, $A=(2,1)$, $B=(-1,0)$
(d) $X=(4,3)$, $A=(2,1)$, $B=(-1,0)$

SOLUTION. (a) $(1,-1)$, $X=A-B$. (b) $(\frac{1}{2},\frac{3}{2})$, $X=\frac{1}{2}A+\frac{3}{2}B$.

(c) $(1,1)$, $X=A+B$. (d) $(3,2)$, $X=3A+2B$.

3. *Find the coordinates of the vector X with respect to the vectors A, B, C.*
(a) $X=(1,0,0)$, $A=(1,1,1)$, $B=(-1,1,0)$, $C=(1,0,-1)$
(b) $X=(1,1,1)$, $A=(0,1,-1)$, $B=(1,1,0)$, $C=(1,0,2)$
(c) $X=(0,0,1)$, $A=(1,1,1)$, $B=(-1,1,0)$, $C=(1,0,-1)$

SOLUTION. (a) $(\frac{1}{3},\frac{-1}{3},\frac{1}{3})$, $X=\frac{1}{3}A+\frac{-1}{3}B+\frac{1}{3}C$.

(b) $(1,0,1)$, $X=A+C$.

(c) $(\frac{1}{3},\frac{-1}{3},\frac{-2}{3})$, $X=\frac{1}{3}A-\frac{1}{3}B-\frac{2}{3}C$.

4. *Let (a,b) and (c,d) be two vectors in the plane. If $ad-bc=0$, show that they are linearly dependent. If $ad-bc\neq 0$, show that they are linearly independent.*

SOLUTION. (i) Suppose that $ad-bc=0$. If one of the vectors (a,b) or (c,d) is O, then both vectors are linearly dependent. Suppose both vectors are non-zero; then we may assume without loss of generality that $c\neq 0$. We contend that $a\neq 0$ and that

$$(a,b)-\frac{a}{c}(c,d)=0 \qquad (*)$$

Indeed, if $a=0$, then $bc=0$; so $b=0$ and $(a,b)=O$ which is a contradiction. Then (*) is true because $b-ad/c=0$.
(ii) Suppose that $ad-bc\neq 0$. Then $x(a,b)+y(c,d)=0$ implies

$$\begin{cases} ax+cy=0 \\ bx+dy=0 \end{cases}$$

Multiplying the first equation by d and subtracting c times the second equation we get $(ad-bc)x=0$; so $x=0$. Hence $cy=0$, and $dy=0$, and the condition $ad-bc\neq 0$ implies $y=0$, so (a,b) and (c,d) are linearly independent.

5. *Consider the vector space of all functions of a variable t. Show that the following pairs of functions are linearly independent.*
(a) $1, t$ *(b)* t, t^2 *(c)* t, t^4 *(d)* e^t, t *(e)* te^t, e^{2t} *(f)* $\sin t, \cos t$
(g) $t, \sin t$ *(h)* $\sin t, \sin 2t$ *(i)* $\cos t, \cos 3t$

SOLUTION. (a) Suppose that $a+bt=0$. Putting $t=0$ and then $t=1$, we find $a=b=0$.

(b) Letting $t=1$ and then $t=-1$ in the equation $at+bt^2=0$, we see that $a=b=0$.

(c) Same as in (b).

(d) Letting $t=0$ and then $t=1$ in the equation $ae^t+bt=0$, we get $a=b=0$.

(e) Let $t=0$ and then $t=1$ in the equation $ate^t+be^{2t}=0$.

(f) Let $t=0$ and then $t=\pi/2$ in the equation $a\cos t+b\sin t=0$.

(h) Let $t=\pi/2$ and then $t=\pi/4$ in the equation $a\sin t+b\sin 2t=0$.

(i) Let $t=\pi/6$ and then $t=0$ in the equation $a\cos t+b\cos 3t=0$.

6. *Consider the vector space of functions defined for $t > 0$. Show that the following pairs of functions are linearly independent.*
(a) $t, 1/t$ *(b)* $e^t, \log t$

SOLUTION. (a) Suppose that $at + b/t = 0$. Let $t = 1$ and $t = 2$ so that $a + b = 0$ and $2a + b/2 = 0$. We conclude at once that $a = b = 0$.

(b) Suppose that $ae^t + b\log t = 0$. Putting $t = 1$, we find $a = 0$, so we see that b must also be 0.

7. *What are the coordinates of the function $3\sin t + 5\cos t = f(t)$ with respect to the basis $\{\sin t, \cos t\}$?*

SOLUTION. $(3, 5)$.

8. *Let D be the derivative d/dt. Let $f(t)$ be as in Exercise 7. What are the coordinates of $Df(t)$ with respect to the basis of Exercise 7?*

SOLUTION. $(-5, 3)$, because $Df(t) = 3\cos t - 5\sin t$.

9. *Let $A_1, \ldots, A_r$ be vectors in $\mathbf{R}^n$ and assume that they are mutually perpendicular (i.e. any two of them are perpendicular), and that none of them is equal to O. Prove that they are linearly independent.*

SOLUTION. Suppose that $a_1A_1 + a_2A_2 + \ldots + a_rA_r = 0$. Then for each i with $1 \le i \le r$ we have

$$0 = A_i \cdot (a_1A_1 + a_2A_2 + \ldots + a_rA_r) = a_1A_i \cdot A_1 + \ldots + a_iA_i \cdot A_i + \ldots + a_rA_i \cdot A_r$$
$$= a_iA_i \cdot A_i.$$

But $A_i \neq O$, so $A_i \cdot A_i > O$ and consequently $a_i = 0$.

10. *Let v, w be elements of a vector space and assume that $v \neq O$. If v, w are linearly independent, show that there is a number a such that $w = av$.*

SOLUTION. If w is zero, let $a = 0$. Assume $w \neq O$; then, since the two vectors v and w are linearly dependent, there exist numbers c and d that are not both zero such that $cv + dw = O$. Hence $cv = -dw$. Since $v \neq O$, we must have $d \neq 0$. Let $a = -c/d$ so that $av = w$.

I, §4 Sums and Direct Sums

1. *Let $V = \mathbf{R}^2$, and let W be the subspace generated by $(2,1)$. Let U be the subspace generated by $(0,1)$. Show that V is the direct sum of W and U. If U' is the subspace generated by $(1,1)$, show that V is also the direct sum of W and U'.*

SOLUTION. (See also Exercise 3.)
(i) Let $(x,y) \in \mathbf{R}^2$. If $a = x/2$ and $b = y - x/2$, then

$$a(2,1) + b(0,1) = (x,y),$$

so $\mathbf{R}^2 = W + U$. If $\alpha(2,1) = \beta(0,1)$, then $\alpha = \beta = 0$. so $W \cap U = \{O\}$. We conclude that $\mathbf{R}^2 = W \oplus U$.
(ii) If $a = x - y$ and $b = 2y - x$, then $a(2,1) + b(1,1) = (x,y)$ so we have $\mathbf{R}^2 = W + U'$. If $\alpha(2,1) = \beta(1,1)$, then $\alpha = \beta = 0$ so $W \cap U' = \{O\}$, and hence $\mathbf{R}^2 = W \oplus U'$.

2. *Let $V = K^3$ for some field K. Let W be the subspace generated by $(1,0,0)$, and let U be the subspace generated by $(1,1,0)$ and $(0,1,1)$. Show that V is the direct sum of W and U.*

SOLUTION. The vector space K^3 has dimension 3, so it is sufficient to show that the three vectors $\{(1,0,0),(1,1,0),(0,1,1)\}$ are linearly independent. Indeed, if $a(1,0,0) + b(1,1,0) + c(0,1,1) = O$, then

$$\begin{cases} a + b = 0 \\ b + c = 0 \\ c = 0 \end{cases}$$

so we must have $a = b = c = 0$ and hence $V = W \oplus U$.

3. *Let A, B be two vectors in $\mathbf{R}^2$ and assume that neither of them is O. If there is no number c such that $cA = B$, show that A, B form a basis for $\mathbf{R}^2$, and that $\mathbf{R}^2$ is the direct sum of the subspaces generated by A and B.*

SOLUTION. The vector space $\mathbf{R}^2$ has dimension 2, so it is sufficient to show that A and B are linearly independent. But suppose not; then there exist numbers a, b that are not 0 such that $aA + bB = 0$ or, equivalently, $aA = -bB$. The number b cannot be 0 because $A \neq O$, so $B = cA$ where $c = -a/b$, which is a contradiction. So $\{A, B\}$ form a basis for $\mathbf{R}^2$.

Now let W_A and W_B be the subspaces generated by A and B, respectively. Since $\{A, B\}$ generates $\mathbf{R}^2$, we have $\mathbf{R}^2 = W_A + W_B$, and the fact that $\{A, B\}$ is a basis implies that any vector $v \in \mathbf{R}^2$ has a unique expression of the form $v = aA + bB$ where $a, b \in \mathbf{R}$. Thus $\mathbf{R}^2 = W_A \oplus W_B$.

4. Prove the last assertion of the section concerning the dimension of $U \times W$. If $\{u_1, \ldots, u_r\}$ is a basis for U and $\{w_1, \ldots, w_s\}$ is a basis for W, what is a basis of $U \times W$?

SOLUTION. We want to show that the dimension of $U \times W$ is $r + s$. Let $A_i = (u_i, O)$ and $B_j = (O, w_j)$. We contend that $S = \{A_i, B_j\}_{\substack{1 \le i \le r \\ 1 \le j \le s}}$ is a basis for $U \times W$. If (u, w) belongs to $U \times W$, then there exist numbers $a_1, \ldots, a_r, b_1, \ldots, b_s$ such that $u = \sum_{i=1}^{r} a_i u_i$ and $w = \sum_{j=1}^{s} b_j w_j$. Then

$$(u, w) = \sum_{i=1}^{r} a_i A_i + \sum_{j=1}^{s} b_j B_j,$$

so S generates $U \times W$. Now we show that the vectors in S are linearly independent. If

$$\sum_{i=1}^{r} a_i A_i + \sum_{j=1}^{s} b_j B_j = (O, O),$$

then $\sum_{i=1}^{r} a_i u_i = O$ and $\sum_{j=1}^{s} b_j w_j = O$ so $a_1 = \ldots = a_r = b_1 = \ldots = b_s = 0$, thereby proving our contention. Hence $\dim(U \times V) = r + s = \dim U + \dim V$.

CHAPTER II

Matrices

II, §1 The Space of Matrices

1. *Let* $A = \begin{pmatrix} 1 & 2 & 3 \\ -1 & 0 & 2 \end{pmatrix}$ *and* $B = \begin{pmatrix} -1 & 5 & -2 \\ 2 & 2 & -1 \end{pmatrix}$. *Find*
$A+B$, $3B$, $-2B$, $A+2B$
$2A-B$, $A-2B$, $B-A$.

SOLUTION. $A+B = \begin{pmatrix} 0 & 7 & 1 \\ 1 & 2 & 1 \end{pmatrix}$, $3B = \begin{pmatrix} -3 & 15 & -6 \\ 6 & 6 & -3 \end{pmatrix}$,
$-2B = \begin{pmatrix} 2 & -10 & 4 \\ -4 & -4 & 2 \end{pmatrix}$, $A+2B = \begin{pmatrix} -1 & 12 & -1 \\ 3 & 4 & 0 \end{pmatrix}$, $2A-B = \begin{pmatrix} 3 & -1 & 8 \\ -4 & -2 & 5 \end{pmatrix}$,
$A-2B = \begin{pmatrix} 3 & -8 & 7 \\ -5 & -4 & 4 \end{pmatrix}$, $B-A = \begin{pmatrix} -2 & 3 & -5 \\ 3 & 2 & -3 \end{pmatrix}$.

2. *Let* $A = \begin{pmatrix} 1 & -1 \\ 2 & 2 \end{pmatrix}$ *and* $B = \begin{pmatrix} -1 & 1 \\ 0 & -3 \end{pmatrix}$. *Find* $A+B$, $3B$, $-2B$, $A+2B$, $A-B$, $B-A$.

SOLUTION. $A+B = \begin{pmatrix} 0 & 0 \\ 2 & -1 \end{pmatrix}$, $3B = \begin{pmatrix} -3 & 3 \\ 0 & -9 \end{pmatrix}$, $-2B = \begin{pmatrix} 2 & -2 \\ 0 & 6 \end{pmatrix}$,
$A+2B = \begin{pmatrix} -1 & 1 \\ 2 & -4 \end{pmatrix}$, $A-B = \begin{pmatrix} 2 & -2 \\ 2 & 5 \end{pmatrix}$, $B-A = \begin{pmatrix} -2 & 2 \\ -2 & -5 \end{pmatrix}$.

3. *In Exercise 1, find* ${}^tA$ *and* ${}^tB$.

SOLUTION. ${}^tA = \begin{pmatrix} 1 & -1 \\ 2 & 0 \\ 3 & 2 \end{pmatrix}$, ${}^tB = \begin{pmatrix} -1 & 2 \\ 5 & 2 \\ -2 & -1 \end{pmatrix}$.

4. *In Exercise 2, find* ^{t}A *and* ^{t}B.

SOLUTION. $^{t}A=\begin{pmatrix}1 & 2\\ -1 & 2\end{pmatrix}$, $^{t}B=\begin{pmatrix}-1 & 0\\ 1 & -3\end{pmatrix}$.

5. *If A, B are arbitrary* $m\times n$ *matrices, show that*

$$^{t}(A+B)={}^{t}A+{}^{t}B.$$

SOLUTION. The matrix $A+B$ is also $m\times n$. Suppose $A=(a_{ij})$ and $B=(b_{ij})$, then the ji-entry of $^{t}(A+B)$ is $a_{ij}+b_{ij}$ and the ji-entries of ^{t}A and ^{t}B are, respectively, a_{ij} and b_{ij}, so we have the formula $^{t}(A+B)={}^{t}A+{}^{t}B$.

6. *If c is a number show that*

$$^{t}(cA)=c\,{}^{t}A.$$

SOLUTION. If $A=(a_{ij})$ and $b_{ji}=a_{ij}$ then by definition we have $^{t}(cA)=(cb_{ji})$; so $c\,{}^{t}A=c(b_{ji})=(cb_{ji})={}^{t}(cA)$.

7. *If* $A=(a_{ij})$ *is a square matrix, then the elements* a_{ii} *are called the diagonal elements. How do the diagonal elements of A and* ^{t}A *differ?*

SOLUTION. The diagonal elements of A and ^{t}A are the same because if $i=j$, then $a_{ij}=a_{ji}$.

8. *Find* $^{t}(A+B)$ *and* $^{t}A+{}^{t}B$ *in Exercise 2.*

SOLUTION. $^{t}(A+B)=\begin{pmatrix}0 & 2\\ 0 & -1\end{pmatrix}={}^{t}A+{}^{t}B$.

9. *Find* $A+{}^{t}A$ *and* $B+{}^{t}B$ *in Exercise 2.*

SOLUTION. $A+{}^{t}A=\begin{pmatrix}2 & 1\\ 1 & 4\end{pmatrix}$, $B+{}^{t}B=\begin{pmatrix}-2 & 1\\ 1 & -6\end{pmatrix}$.

10. *Show that for any square matrix A, the matrix* $A+{}^{t}A$ *is symmetric.*

SOLUTION. If $A=(a_{ij})$, then the *ij*-entry and *ji*-entry of the matrix $A+{}^tA$ are $a_{ij}+a_{ji}$ and $a_{ji}+a_{ij}$, respectively, so $A+{}^tA$ is symmetric.

11. *Write down the row vectors and column vectors of the matrices A, B in Exercise 1.*

SOLUTION.
Matrix A. 1st row $=(1\ \ 2\ \ 3)$, and 2nd row $=(-1\ \ 0\ \ 2)$.

$$1^{\text{st}} \text{ column } =\begin{pmatrix}1\\-1\end{pmatrix},\ 2^{\text{nd}} \text{ column } =\begin{pmatrix}2\\0\end{pmatrix},\ 3^{\text{rd}} \text{ column } =\begin{pmatrix}3\\2\end{pmatrix}.$$

Matrix B. 1st row $=(-1\ \ 5\ \ -2)$, and 2nd row $=(2\ \ 2\ \ -1)$.

$$1^{\text{st}} \text{ column } =\begin{pmatrix}-1\\2\end{pmatrix},\ 2^{\text{nd}} \text{ column } =\begin{pmatrix}5\\2\end{pmatrix},3^{\text{rd}} \text{ column } =\begin{pmatrix}-2\\-1\end{pmatrix}.$$

12. *Write down the row vectors and column vectors of the matrices A, B in Exercise 2.*

SOLUTION.
Matrix A. 1st row $=(1\ \ -1)$, and 2nd row $=(2\ \ 2)$.

$$1^{\text{st}} \text{ column } =\begin{pmatrix}1\\2\end{pmatrix},\ 2^{\text{nd}} \text{ column } =\begin{pmatrix}-1\\2\end{pmatrix}.$$

Matrix B. 1st row $=(-1\ \ 1)$, and 2nd row $=(0\ \ -3)$.

$$1^{\text{st}} \text{ column } =\begin{pmatrix}-1\\0\end{pmatrix},\ 2^{\text{nd}} \text{ column } =\begin{pmatrix}1\\-3\end{pmatrix}.$$

II, §1 The Space of Matrices

In this section we let E_{ij} be the matrix with all entries 0 except the *ij*-entry, which is equal to 1. We call these matrices the elementary matrices.

1. *What is the dimension of the space of 2×2 matrices? Give a basis for this space.*

SOLUTION. The space of 2×2 matrices has dimension 4. The matrices

$$E_{1,1}=\begin{pmatrix}1&0\\0&0\end{pmatrix},\quad E_{1,2}=\begin{pmatrix}0&1\\0&0\end{pmatrix},\quad E_{2,1}=\begin{pmatrix}0&0\\1&0\end{pmatrix},\quad E_{2,2}=\begin{pmatrix}0&0\\0&1\end{pmatrix}$$

clearly form a basis for the space of 2×2 matrices. See Exercise 2.

2. *What is the dimension of the space of $m\times n$ matrices? Give a basis for this space.*

SOLUTION. The space of $m\times n$ matrices has dimension mn. The set $S=\{E_{ij}\}_{\substack{1\le i\le m\\ 1\le j\le n}}$ is a basis for the space $\mathrm{Mat}_{m\times n}(K)$. Indeed, if (a_{ij}) is an $m\times n$ matrix, then

$$(a_{ij})=\sum_{j=1}^{n}\sum_{i=1}^{m}a_{ij}E_{ij},$$

so S generates $\mathrm{Mat}_{m\times n}(K)$. The vectors of S are linearly independent because if

$$\sum_{j=1}^{n}\sum_{i=1}^{m}c_{ij}E_{ij}=O,$$

then $(c_{ij})=O$; thus $c_{ij}=0$. Clearly, S has mn elements

3. *What is the dimension of the space of $n\times n$ matrices all of whose components are 0 except possibly the diagonal components?*

SOLUTION. The dimension of the space considered is n, and the set $S=\{E_{ii}\}_{1\le i\le n}$ is a basis.

4. *What is the dimension of the space of $n\times n$ matrices which are upper triangular, i.e. of the following type:*

$$\begin{pmatrix} a_{11} & a_{12} & \dots & a_{1n}\\ 0 & a_{22} & \dots & a_{2n}\\ \vdots & \vdots & & \vdots\\ 0 & 0 & \dots & a_{nn}\end{pmatrix}?$$

SOLUTION. The dimension of the space W of $n\times n$ upper triangular matrices is $\frac{n(n+1)}{2}$, because the set $S=\{E_{ij}\}_{1\le i\le j\le n}$ is a basis for W. Actually, S consists of all elementary matrices that are upper triangular. If $(a_{ij})\in W$, then

$$(a_{ij})=\sum_{1\le i\le j\le n}a_{ij}E_{ij},$$

and if $\sum_{1\le i\le j\le n} c_{ij}E_{ij} = O$, then we must have $c_{ij} = 0$. The set S has $\dfrac{n(n+1)}{2}$ elements. Indeed, when $i = 1$, we have n vectors, namely, $E_{1,1}, E_{1,2},\ldots, E_{1,n}$. When $i = 2$ then we have $n-1$ vectors, namely, $E_{2,2}, E_{2,3},\ldots, E_{2,n}$. But

$$n+(n-1)+(n-2)+\ldots+1=\frac{n(n+1)}{2}.$$

Note: We can compute the cardinal p of S in another way. If we subtract the diagonal elements from S we get all of the elements strictly above the diagonal. By symmetry we have the same number of elements strictly below the diagonal, so we multiply $p-n$ by 2. Finally, adding the diagonal elements we see that $2(p-n)+n=n^2$. Now we solve for p.

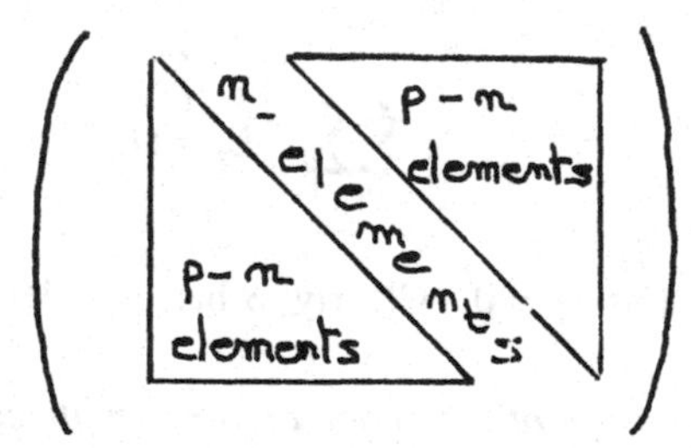

5. *What is the dimension of the space of symmetric 2×2 (i.e. 2×2 matrices A such that $A={}^tA$)? Exhibit a basis for this space.*

SOLUTION. The space in question has dimension 3, and a basis for this space is given by

$$E_{1,1}=\begin{pmatrix}1&0\\0&0\end{pmatrix},\quad E_{1,2}+E_{2,1}=\begin{pmatrix}0&1\\1&0\end{pmatrix},\quad E_{2,2}\begin{pmatrix}0&0\\0&1\end{pmatrix}$$

6. *More generally what is the dimension of the space of symmetric $n\times n$ matrices? What is a basis for this space?*

SOLUTION. The dimension of the space of symmetric $n\times n$ matrices is $\dfrac{n(n+1)}{2}$. A basis for this space is $\{E_{ij}+E_{ji}\}_{1\le i<j\le n}\cup\{E_{ii}\}_{1\le i\le n}$. If k is the cardinal of $\{E_{ij}+E_{ji}\}_{1\le i<j\le n}$, then k is also the number of strictly upper triangular elementary matrices. Therefore, arguing as in Exercise 4, we see that $2k+n=n^2$, so that $k+n=\dfrac{n(n+1)}{2}$.

7. *What is the dimension of the space of diagonal $n \times n$ matrices? What is a basis for this space?*

SOLUTION. The dimension of the space of diagonal $n \times n$ matrices is n because a basis for this space is simply $\{E_{ii}\}_{1 \le i \le n}$.

8. *Let V be a subspace of* $\mathbf{R}^2$. *What are the possible dimensions for V?*

SOLUTION. All the possible dimensions for V are 0, 1, or 2. Theorem 3.7 of Chapter I implies that V can have dimension 0, 1, or 2. An example for each case would be O for dimension 0, a line passing through the origin for dimension 1, and $\mathbf{R}^2$ itself for dimension 2.

9. *Let V be a subspace of* $\mathbf{R}^3$. *What are the possible dimensions for V?*

SOLUTION. All the possible dimensions for V are 0, 1, 2, or 3. An example for each case would be O for dimension 0, a line passing through the origin for dimension 1, a plane passing through the origin for dimension 2, and $\mathbf{R}^3$ itself for dimension 3.

II, §2 Linear Equations

1. *Let* $(**)$ *be a system of homogeneous linear equations in a field K, and assume that* $m = n$. *Assume also that the column vectors of coefficients are linearly independent. Show that the only solution is the trivial solution.*

SOLUTION. If the vectors A^i are linearly independent, then $x_1 A^1 + \ldots + x_n A^n = O$ if and only if $x_1 = \ldots = x_n = 0$.

2. *Let* $(**)$ *be a system of homogeneous linear equations in a field K, in n unknowns. Show that the set of solutions* $X = (x_1, \ldots, x_n)$ *is a vector space over K.*

SOLUTION. Since the system is homogeneous, the vector $O = (0, 0, \ldots, 0)$ is a solution of the system. Because the system is linear, we see that if $(x_1, \ldots, x_n)$ and $(y_1, \ldots, y_n)$ are solutions, then $(x_1 + y_1, \ldots, x_n + y_n)$ and $(cx_1, \ldots, cx_n)$ are also solutions.

3. *Let* $A^1, \ldots, A^n$ *be column vectors of size m. Assume that they have coefficients in* $\mathbf{R}$, *and that they are linearly independent over* $\mathbf{R}$. *Show that they are linearly independent over* $\mathbf{C}$.

SOLUTION. Let $c_1,\ldots,c_n$ be complex numbers such that $c_1A^1+\ldots+c_nA^n=O$. We can write $c_k=x_k+y_ki$ so that the preceding equation becomes

$$x_1A^1+\ldots+x_nA^n+i\left(y_1A^1+\ldots+y_nA^n\right)=O \qquad (*)$$

But a complex number is zero if and only if its real and imaginary parts are 0, so writing down the coordinates of each column vector we see that (*) implies the two systems $x_1A^1+\ldots+x_nA^n=O$ and $y_1A^1+\ldots+y_nA^n=O$. The column vectors are linearly independent over **R**, so we get $c_1=\ldots=c_n=0$.

4. *Let (**) be a system of homogeneous linear equations with coefficients in **R**. If this system has a non-trivial solution in **C**, show that it has a non-trivial solution in **R**.*

SOLUTION. Suppose that the system only has the trivial solution in **R**. Then we see that the column vectors are linearly independent over **R**, so Exercise 3 implies that the column vectors are linearly independent over **C**, and consequently the system has only the trivial solution in **C**, which is a contradiction.

II, §3 Multiplication of Matrices

1. *Let I be the unit $n\times n$ matrix. Let A be an $n\times r$ matrix. What is IA? If A is an $m\times n$ matrix what is AI?*

SOLUTION. We have $IA=AI=A$ because if $I=\left(\delta_{ij}\right)$, where $\delta_{ij}=1$ if $i=j$ and $\delta_{ij}=0$ if $i\neq j$ and $A=\left(a_{ij}\right)_{\substack{1\le i\le n\\ 1\le j\le r}}$, then the lk-entry of IA and AI is a_{lk}. Indeed, $\sum_{j=1}^{n}\delta_{lj}a_{jk}=\sum_{j=1}^{n}a_{lj}\delta_{jk}=a_{lk}$.

2. *Let O be the matrix all whose coordinates are 0. Let A be a matrix of a size such that the product AO is defined. What is AO?*

SOLUTION. Clearly $AO=O$.

3. *In each one of the following cases, find $(AB)C$ and $A(BC)$.*

(a) $A=\begin{pmatrix}2&1\\3&1\end{pmatrix}$, $B=\begin{pmatrix}-1&1\\1&0\end{pmatrix}$, $C=\begin{pmatrix}1&4\\2&3\end{pmatrix}$

(b) $A=\begin{pmatrix}2 & 1 & -1\\ 3 & 1 & 2\end{pmatrix}$, $B=\begin{pmatrix}1 & 1\\ 2 & 0\\ 3 & -1\end{pmatrix}$, $C=\begin{pmatrix}1\\ 3\end{pmatrix}$

(c) $A=\begin{pmatrix}2 & 4 & 1\\ 3 & 0 & -1\end{pmatrix}$, $B=\begin{pmatrix}1 & 1 & 0\\ 2 & 1 & -1\\ 3 & 1 & 5\end{pmatrix}$, $C=\begin{pmatrix}1 & 2\\ 3 & 1\\ -1 & 4\end{pmatrix}$

SOLUTION.

(a) $(AB)C=\begin{pmatrix}-1 & 2\\ -2 & 3\end{pmatrix}\begin{pmatrix}1 & 4\\ 2 & 3\end{pmatrix}=\begin{pmatrix}3 & 2\\ 4 & 1\end{pmatrix}$, $A(BC)=\begin{pmatrix}2 & 1\\ 3 & 1\end{pmatrix}\begin{pmatrix}1 & -1\\ 1 & 4\end{pmatrix}=\begin{pmatrix}3 & 2\\ 4 & 1\end{pmatrix}$.

(b) $(AB)C=\begin{pmatrix}1 & 3\\ 11 & 1\end{pmatrix}\begin{pmatrix}1\\ 3\end{pmatrix}=\begin{pmatrix}10\\ 14\end{pmatrix}$, $A(BC)=\begin{pmatrix}2 & 1 & -1\\ 3 & 1 & 2\end{pmatrix}\begin{pmatrix}4\\ 2\\ 0\end{pmatrix}=\begin{pmatrix}10\\ 14\end{pmatrix}$.

(c) $(AB)C=\begin{pmatrix}13 & 7 & 1\\ 0 & 2 & -5\end{pmatrix}\begin{pmatrix}1 & 2\\ 3 & 1\\ -1 & 4\end{pmatrix}=\begin{pmatrix}33 & 37\\ 11 & -18\end{pmatrix}$,

$$A(BC)=\begin{pmatrix}2 & 4 & 1\\ 3 & 0 & -1\end{pmatrix}\begin{pmatrix}4 & 3\\ 6 & 1\\ 1 & 27\end{pmatrix}=\begin{pmatrix}33 & 37\\ 11 & -18\end{pmatrix}.$$

4. *Let A, B be square matrices of the same size, and assume that $AB=BA$. Show that $(A+B)^2=A^2+2AB+B^2$, and*

$$(A+B)(A-B)=A^2-B^2,$$

using the properties of matrices stated in Theorem 3.1.

SOLUTION. We have

$$\begin{aligned}(A+B)(A+B)=(A+B)A+(A+B)B&=A^2+BA+AB+B^2\\ &=A^2+2AB+B^2\end{aligned}$$

and

$$(A+B)(A-B)=(A+B)A-(A+B)B=A^2-B^2.$$

5. *Let* $A=\begin{pmatrix}1 & 2\\ 3 & -1\end{pmatrix}$, $B=\begin{pmatrix}2 & 0\\ 1 & 1\end{pmatrix}$. *Find AB and BA.*

SOLUTION. Doing the computation we find that

$$AB = \begin{pmatrix} 4 & 2 \\ 5 & -1 \end{pmatrix} \quad \text{and} \quad BA = \begin{pmatrix} 2 & 4 \\ 4 & 1 \end{pmatrix}.$$

6. *Let* $C = \begin{pmatrix} 7 & 0 \\ 0 & 7 \end{pmatrix}$. *Let A, B be as in Exercise 5. Find CA, AC, CB and BC. State the general rule including this exercise as a special case.*

SOLUTION. Note that $C = 7I$. The computation shows that

$$CA = AC = \begin{pmatrix} 7 & 14 \\ 21 & -7 \end{pmatrix} \quad \text{and} \quad CB = BC = \begin{pmatrix} 14 & 0 \\ 7 & 7 \end{pmatrix}.$$

The rule is that in general we **do not** always have $AB = BA$.

7. *Let* $X = (1,0,0)$ *and let* $A = \begin{pmatrix} 3 & 1 & 5 \\ 2 & 0 & 1 \\ 1 & 1 & 7 \end{pmatrix}$. *What is XA?*

SOLUTION. We find that $XA = (3 \ \ 1 \ \ 5)$.

8. *Let* $X = (0,1,0)$, *and let A be an arbitrary* 3×3 *matrix. How would you describe XA? What if* $X = (0,0,1)$? *Generalize to similar statements concerning matrices, and their products with unit vectors.*

SOLUTION. We solve the general case. Consider an $n \times n$ matrix, say $A = (a_{ij})$. Let $X_k = (0,\ldots,0,1,0,\ldots,0)$ be the row vector with zeros everywhere except 1 at the k-entry. Then we see that

$$X_k A = (0,,\ldots,0,1,0,\ldots,0) \begin{pmatrix} a_{11} & \cdots & & a_{1n} \\ \vdots & & & \vdots \\ a_{k1} & a_{k2} & \cdots & a_{kn} \\ \vdots & & & \vdots \\ a_{n1} & \cdots & & a_{nn} \end{pmatrix} = (a_{k1} \ \ a_{k2} \ \ \ldots \ \ a_{kn})$$

so $X_k A$ equals the kth row of A, namely A_k.

9. *Let A, B be the matrices of Exercise 3(a). Verify by computation that* ${}^t(AB) = {}^tB\,{}^tA$. *Do the same 3(b) and 3(c). Prove the same rule for any two* *matrices A, B (which can be multiplied). If A, B, C are matrices which can be multiplied, show that* ${}^t(ABC)={}^tC\,{}^tB\,{}^tA$.

SOLUTION. (i) For the matrices of Exercise 3(a) we find

$${}^tB\,{}^tA=\begin{pmatrix}-1&1\\1&0\end{pmatrix}\begin{pmatrix}2&3\\1&1\end{pmatrix}=\begin{pmatrix}-1&-2\\2&3\end{pmatrix}={}^t(AB).$$

For the matrices of Exercise 3(b) we find

$${}^tB\,{}^tA=\begin{pmatrix}1&2&3\\1&0&-1\end{pmatrix}\begin{pmatrix}2&3\\1&1\\-1&2\end{pmatrix}=\begin{pmatrix}1&11\\3&1\end{pmatrix}={}^t(AB).$$

For the matrices of Exercise 3(c) we find

$${}^tB\,{}^tA=\begin{pmatrix}1&2&3\\1&1&1\\0&-1&5\end{pmatrix}\begin{pmatrix}2&3\\4&0\\1&-1\end{pmatrix}=\begin{pmatrix}13&0\\7&2\\1&-5\end{pmatrix}={}^t(AB).$$

(ii) In general, suppose that $A=(a_{ij})$ is an $m\times n$ and that $B=(b_{kl})$ is an $n\times p$ matrix. The rs-entry of the matrix ${}^t(AB)$ is the sr-entry of the matrix AB, namely,

$$\sum_{j=1}^{n}a_{sj}b_{jr}\,.$$

The rs-entry of the product ${}^tB\,{}^tA$ is given by

$$\begin{pmatrix}b_{1r}&\cdots&b_{nr}\end{pmatrix}\begin{pmatrix}a_{s1}\\ \vdots\\ a_{sn}\end{pmatrix}=\sum_{j=1}^{n}b_{jr}a_{sj}\,,$$

so ${}^t(AB)={}^tB\,{}^tA$.

(iii) Finally, the formula ${}^t(ABC)={}^tC\,{}^tB\,{}^tA$ holds because

$${}^t(ABC)={}^t((AB)C)={}^tC\,{}^t(AB)={}^tC\,{}^tB\,{}^tA.$$

10. *Let M be an* $n\times n$ *matrix such that* ${}^tM=M$. *Given two row vectors in n-space, say A and B define* $\langle A,B\rangle$ *to be* $AM\,{}^tB$. *(Identify a* 1×1 *matrix with a number.) Show that the conditions of a scalar product are satisfied , ex-*

cept possibly the condition concerning positivity. Give an example of a matrix M and vectors A, B such that $AM'B$ is negative (taking $n=2$).

SOLUTION. For all square matrices we have ${}^t({}^tA) = A$. This result combined with the formula of Exercise 9 and the fact that $M={}^tM$ implies

$$A M^t B = {}^t(B M^t A) = B M^t A,$$

the last equality holding because BM^tA is a 1×1 matrix. Thus **SP 1** holds because $\langle A, B\rangle = \langle B, A\rangle$.

For **SP 2**, note that

$$\langle A, B+C\rangle = AM^t(B+C) = AM({}^tB + {}^tC) = AM^tB + AM^tC = \langle A, B\rangle + \langle A, C\rangle.$$

Finally, **SP 3** holds because if c is a number, we have

$$\langle cA, B\rangle = cAM^tB = c\langle A, B\rangle \quad \text{and} \quad \langle A, cB\rangle = AM^t(cB) = c\langle A, B\rangle.$$

Suppose $n=2$. Here are two examples which illustrate that positivity need not hold:
(i) If $M = O$, then $\langle A, A\rangle = 0$ for all A.
(ii) If $M = \begin{pmatrix} 0 & 1 \\ 1 & 0 \end{pmatrix}$ and $A = (1 \ \ 0)$, then $\langle A, A\rangle = (1 \ \ 0)\begin{pmatrix} 0 \\ 1 \end{pmatrix} = 0$.

11. *(a) Let A be the matrix* $\begin{pmatrix} 0 & 1 & 1 \\ 0 & 0 & 1 \\ 0 & 0 & 0 \end{pmatrix}$. *Find* A^2, A^3. *Generalize to* 4×4 *matrices.*

(b) Let A be the matrix $\begin{pmatrix} 1 & 1 & 1 \\ 0 & 1 & 1 \\ 0 & 0 & 1 \end{pmatrix}$. *Compute* A^2, A^3, A^4.

SOLUTION. (a) The computations show that

$$A^2 = \begin{pmatrix} 0 & 0 & 1 \\ 0 & 0 & 0 \\ 0 & 0 & 0 \end{pmatrix} \quad \text{and} \quad A^3 = \begin{pmatrix} 0 & 0 & 0 \\ 0 & 0 & 0 \\ 0 & 0 & 0 \end{pmatrix}.$$

For the general theorem, see Exercise 35.

(b) We have

$$A^2 = \begin{pmatrix} 1 & 2 & 3 \\ 0 & 1 & 2 \\ 0 & 0 & 1 \end{pmatrix}, \quad A^3 = \begin{pmatrix} 1 & 3 & 6 \\ 0 & 1 & 3 \\ 0 & 0 & 1 \end{pmatrix}, \text{ and } A^4 = \begin{pmatrix} 1 & 4 & 10 \\ 0 & 1 & 4 \\ 0 & 0 & 1 \end{pmatrix}.$$

12. *Let X be the indicated column vector, and A the indicated matrix. Find AX as a column vector.*

(a) $X = \begin{pmatrix} 3 \\ 2 \\ 1 \end{pmatrix}$, $A = \begin{pmatrix} 1 & 0 & 1 \\ 2 & 0 & 1 \\ 2 & 0 & -1 \end{pmatrix}$

(b) $X = \begin{pmatrix} 1 \\ 1 \\ 0 \end{pmatrix}$, $A = \begin{pmatrix} 2 & 1 & 5 \\ 0 & 1 & 1 \end{pmatrix}$

(c) $X = \begin{pmatrix} x_1 \\ x_2 \\ x_3 \end{pmatrix}$, $A = \begin{pmatrix} 0 & 1 & 0 \\ 0 & 0 & 0 \end{pmatrix}$

(d) $X = \begin{pmatrix} x_1 \\ x_2 \\ x_3 \end{pmatrix}$, $A = \begin{pmatrix} 0 & 0 & 0 \\ 1 & 0 & 0 \end{pmatrix}$

SOLUTION.

(a) $\begin{pmatrix} 4 \\ 7 \\ 5 \end{pmatrix}$ (b) $\begin{pmatrix} 3 \\ 1 \end{pmatrix}$ (c) $\begin{pmatrix} x_2 \\ 0 \end{pmatrix}$ (d) $\begin{pmatrix} 0 \\ x_1 \end{pmatrix}$.

13. *Let* $A = \begin{pmatrix} 2 & 1 & 3 \\ 4 & 1 & 5 \end{pmatrix}$. *Find AX for each of the following values of X.*

(a) $X = \begin{pmatrix} 1 \\ 0 \\ 0 \end{pmatrix}$ *(b)* $X = \begin{pmatrix} 0 \\ 1 \\ 1 \end{pmatrix}$ *(c)* $X = \begin{pmatrix} 0 \\ 0 \\ 1 \end{pmatrix}$

SOLUTION. (a) $\begin{pmatrix} 2 \\ 4 \end{pmatrix}$ (b) $\begin{pmatrix} 4 \\ 6 \end{pmatrix}$ (c) $\begin{pmatrix} 3 \\ 5 \end{pmatrix}$.

14. *Let* $A = \begin{pmatrix} 3 & 7 & 5 \\ 1 & -1 & 4 \\ 2 & 1 & 8 \end{pmatrix}$. *Find AX for each of the values of X given in Exercise 13.*

SOLUTION. (a) $\begin{pmatrix} 3 \\ 1 \\ 2 \end{pmatrix}$ (b) $\begin{pmatrix} 12 \\ 3 \\ 9 \end{pmatrix}$ (c) $\begin{pmatrix} 5 \\ 4 \\ 8 \end{pmatrix}$.

15. *Let*

$$X = \begin{pmatrix} 0 \\ 1 \\ 0 \\ 0 \end{pmatrix} \quad \textit{and} \quad A = \begin{pmatrix} a_{11} & \dots & a_{14} \\ \vdots & & \vdots \\ a_{m1} & \dots & a_{m4} \end{pmatrix}.$$

What is AX?

SOLUTION. A computation shows that AX is equal to the second column of A.

16. *Let X be a column vector having all its components equal to 0 except the i-th component which is equal to 1. Let A be an arbitrary matrix, whose size is such that we can form the product AX. What is AX?*

SOLUTION. The product AX equals the ith column of A because

$$AX = \begin{pmatrix} a_{11} & \dots & a_{1i} & \dots & a_{1n} \\ \vdots & & \vdots & & \vdots \\ a_{m1} & \dots & a_{mi} & \dots & a_{mn} \end{pmatrix} \begin{pmatrix} 0 \\ \vdots \\ 1 \\ \vdots \\ 0 \end{pmatrix} = \begin{pmatrix} a_{1i} \\ a_{2i} \\ \\ a_{mi} \end{pmatrix}.$$

17. *Let* $A = (a_{ij})$, $i = 1,\dots,m$ *and* $j = 1,\dots,n$ *be an* $m \times n$ *matrix. Let* $B = (b_{jk})$, $j = 1,\dots,n$ *and* $k = 1,\dots,s$, *be an* $n \times s$ *matrix. Let* $AB = C$. *Show that the k-th column* C^k *can be written* $C^k = b_{1k}A^1 + \dots + b_{nk}A^n$. *(This will be useful in finding the determinant of a product).*

SOLUTION. The p-entry of C^k is given by $\sum_{j=1}^{n} a_{pj}b_{jk}$. The p-entry of the sum $b_{1k}A^1 + \dots + b_{nk}A^n$ is

$$b_{1k}a_{p1} + \dots + b_{nk}a_{pn} = \sum_{j=1}^{n} a_{pj}b_{jk},$$

so $C^k = b_{1k}A^1 + \ldots + b_{nk}A^n$, as was to be shown.

18. *Let A be a square matrix.*
(a) If $A^2 = O$ show that $I - A$ is invertible.
(b) If $A^3 = O$ show that $I - A$ is invertible.
(c) In general, if $A^n = O$ for some positive integer n show that $I - A$ is invertible.
(d) Suppose that $A^2 + 2A + I = O$. Show that A is invertible.
(e) Suppose that $A^2 - A + I = O$. Show that A is invertible.

SOLUTION. (a), (b) and (c) For all positive integers n we have

$$(I - A)(I + A + A^2 + \ldots + A^{n-1}) = I + A + A^2 + \ldots + A^{n-1} - A - \ldots - A^{n-1} - A^n,$$

so $(I - A)(I + A + \ldots + A^{n-1}) = I - A^n$ and $(I + A + \ldots + A^{n-1})(I - A) = I - A^n$. Thus $I - A$ is invertible, and its inverse is given by $I + A + \ldots + A^{n-1}$.

(d) The result follows from the fact that

$$A(-A - 2I) = (-A - 2I)A = -A^2 - 2A = I.$$

(e) It suffices to see that

$$A(-A^2 + I) = (-A^2 + I)A = -A^3 + A = I.$$

19. *Let a, b be numbers, and let* $A = \begin{pmatrix} 1 & a \\ 0 & 1 \end{pmatrix}$ *and* $B = \begin{pmatrix} 1 & b \\ 0 & 1 \end{pmatrix}$. *What is AB? What is A^n where n is a positive integer?*

SOLUTION. A simple computation shows that $AB = \begin{pmatrix} 1 & b+a \\ 0 & 1 \end{pmatrix}$. By induction we prove that $A^n = \begin{pmatrix} 1 & na \\ 0 & 1 \end{pmatrix}$. Clearly the result is true for $n = 1$. If the formula is true for some positive integer n, then we have

$$A^{n+1} = AA^n = A^nA = \begin{pmatrix} 1 & na \\ 0 & 1 \end{pmatrix}\begin{pmatrix} 1 & a \\ 0 & 1 \end{pmatrix} = \begin{pmatrix} 1 & (n+1)a \\ 0 & 1 \end{pmatrix}.$$

20. *Show that the matrix A in Exercise 19 has an inverse. What is this inverse?*

SOLUTION. The matrix in Exercise 19 has an inverse, namely, $A^{-1} = \begin{pmatrix} 1 & -a \\ 0 & 1 \end{pmatrix}$.

21. *Show that if A, B are $n \times n$ matrices which have inverses, then AB has an inverse.*

SOLUTION. The inverse of AB is $B^{-1}A^{-1}$ because

$$(AB)(B^{-1}A^{-1}) = A(BB^{-1})A^{-1} = AA^{-1} = I$$

and

$$(B^{-1}A^{-1})(AB) = B^{-1}(A^{-1}A)B = B^{-1}B = I.$$

22. *Determine all 2×2 matrices A such that $A^2 = O$.*

SOLUTION. Suppose

$$\begin{pmatrix} a & b \\ c & d \end{pmatrix}\begin{pmatrix} a & b \\ c & d \end{pmatrix} = \begin{pmatrix} a^2 + bc & ab + bd \\ ac + cd & cb + d^2 \end{pmatrix},$$

Then we find $a^2 = d^2 = -bc$ and $b(a+d) = c(a+d) = 0$. $(*)$

Case 1. If $a = -d$, then we see that $-bc = a^2$. It is a trivial computation to verify that the matrices of the form

$$\begin{pmatrix} a & b \\ c & -a \end{pmatrix}$$

where $-bc = a^2$, are solutions of the equation $A^2 = O$.

Case 2. Suppose that $a = d$ and assume that $a \neq 0$; otherwise, we are in case 1. From $(*)$ we see that both b and c are 0, so we must have $a = 0$ and therefore we are back in case 1.

Hence the matrices solution of $A^2 = O$ are the matrices of the form $\begin{pmatrix} a & b \\ c & -a \end{pmatrix}$ where $-bc = a^2$.

23. *Let* $A = \begin{pmatrix} \cos\theta & -\sin\theta \\ \sin\theta & \cos\theta \end{pmatrix}$. *Show that* $A^2 = \begin{pmatrix} \cos 2\theta & -\sin 2\theta \\ \sin 2\theta & \cos 2\theta \end{pmatrix}$. *Determine A^n by induction for any positive integer n.*

SOLUTION. Induction proves that $A^n = \begin{pmatrix} \cos n\theta & -\sin n\theta \\ \sin n\theta & \cos n\theta \end{pmatrix}$. Indeed, the trigonometric formulas

$$\cos a \cos b - \sin a \sin b = \cos(a+b)$$

and

$$\cos a \ \sin b + \sin a \cos b = \sin(a+b)$$

imply

$$\begin{pmatrix} \cos\theta & -\sin\theta \\ \sin\theta & \cos\theta \end{pmatrix}\begin{pmatrix} \cos n\theta & -\sin n\theta \\ \sin n\theta & \cos n\theta \end{pmatrix} = \begin{pmatrix} \cos(n+1)\theta & -\sin(n+1)\theta \\ \sin(n+1)\theta & \cos(n+1)\theta \end{pmatrix}.$$

24. *Find a* 2×2 *matrix A such that* $A^2 = -I = \begin{pmatrix} -1 & 0 \\ 0 & -1 \end{pmatrix}$.

SOLUTION. From the solution of Exercise 22 we see that we can choose $a=-d=1$, $b=2$, and $c=-1$. Indeed,

$$\begin{pmatrix} 1 & 2 \\ -1 & -1 \end{pmatrix}\begin{pmatrix} 1 & 2 \\ -1 & -1 \end{pmatrix} = \begin{pmatrix} -1 & 0 \\ 0 & -1 \end{pmatrix}.$$

25. *Let A be an* $n\times n$ *matrix. Define the trace of A to be the sum of the diagonal elements. Thus if* $A=(a_{ij})$, *then* $\operatorname{tr}(A) = \sum_{i=1}^{n} a_{ii}$. *Compute the trace of the following matrices:*

(a) $\begin{pmatrix} 1 & 7 & 3 \\ -1 & 5 & 2 \\ 2 & 3 & -4 \end{pmatrix}$ *(b)* $\begin{pmatrix} 3 & -2 & 4 \\ 1 & 4 & 1 \\ -7 & -3 & -3 \end{pmatrix}$ *(c)* $\begin{pmatrix} -2 & 1 & 1 \\ 3 & 4 & 4 \\ -5 & 2 & 6 \end{pmatrix}$

SOLUTION. (a) 2 (b) 4 (c) 8.

26. *Let A, B be the indicated matrices. Show that* $\operatorname{tr}(AB) = \operatorname{tr}(BA)$.

(a) $A = \begin{pmatrix} 1 & -1 & 1 \\ 2 & 4 & 1 \\ 3 & 0 & 1 \end{pmatrix}$, $B = \begin{pmatrix} 3 & 1 & 2 \\ 1 & 1 & 0 \\ -1 & 2 & 1 \end{pmatrix}$

$$(b)\ A=\begin{pmatrix}1 & 7 & 3\\ -1 & 5 & 2\\ 2 & 3 & -4\end{pmatrix},\quad B=\begin{pmatrix}3 & -2 & 4\\ 1 & 4 & 1\\ -7 & -3 & 2\end{pmatrix}$$

SOLUTION. (a) $\mathrm{tr}(AB)=\mathrm{tr}(BA)=16$ (b) $\mathrm{tr}(AB)=\mathrm{tr}(BA)=8$.

27. *Prove in general that if A, B are square $n\times n$ matrices, then*

$$\mathrm{tr}(AB)=\mathrm{tr}(BA).$$

SOLUTION. Suppose that $A=(a_{ij})$, $B=(b_{ij})$, $AB=(c_{ij})$, and $BA=(d_{ij})$. Then

$$\begin{aligned}\mathrm{tr}(AB)=\sum_{k=1}^{n}c_{kk}&=\sum_{k=1}^{n}\sum_{p=1}^{n}a_{kp}b_{pk}\\ &=\sum_{p=1}^{n}\sum_{k=1}^{n}b_{pk}a_{kp}=\sum_{p=1}^{n}d_{pp}=\mathrm{tr}(BA).\end{aligned}$$

28. *For any square matrix A, show that* $\mathrm{tr}(A)=\mathrm{tr}({}^{t}A)$.

SOLUTION. Taking the transpose leaves the diagonal unchanged, so $\mathrm{tr}(A)=\mathrm{tr}({}^{t}A)$.

29. *Let* $A=\begin{pmatrix}1 & 0 & 0\\ 0 & 2 & 0\\ 0 & 0 & 3\end{pmatrix}$. *Find* A^2, A^3, A^4.

SOLUTION. We find $A^k=\begin{pmatrix}1 & 0 & 0\\ 0 & 2^k & 0\\ 0 & 0 & 3^k\end{pmatrix}$ for $k=2$, 3, and 4. See Exercise 30.

30. *Let A be a diagonal matrix, with diagonal elements $a_1,\ldots,a_n$. What is A^2, A^3, A^k for any positive integer k?*

SOLUTION. We prove by induction that

$$A^k = \begin{pmatrix} a_1^k & 0 & \dots & 0 \\ 0 & a_2^k & \dots & 0 \\ \vdots & \vdots & & \vdots \\ 0 & 0 & \dots & a_n^k \end{pmatrix}.$$

The formula is true when $k=1$. Suppose that the formula is true for some positive integer k; then the rules of multiplication of matrices imply that

$$A^{k+1} = AA^k = \begin{pmatrix} a_1 & 0 & \dots & 0 \\ 0 & a_2 & \dots & 0 \\ \vdots & \vdots & & \vdots \\ 0 & 0 & \dots & a_n \end{pmatrix}\begin{pmatrix} a_1^k & 0 & \dots & 0 \\ 0 & a_2^k & \dots & 0 \\ \vdots & \vdots & & \vdots \\ 0 & 0 & \dots & a_n^k \end{pmatrix} = \begin{pmatrix} a_1^{k+1} & 0 & \dots & 0 \\ 0 & a_2^{k+1} & \dots & 0 \\ \vdots & \vdots & & \vdots \\ 0 & 0 & \dots & a_n^{k+1} \end{pmatrix}.$$

31. *Let* $A = \begin{pmatrix} 0 & 1 & 6 \\ 0 & 0 & 4 \\ 0 & 0 & 0 \end{pmatrix}$. *Find* A^3.

SOLUTION. We find that $A^3 = O$.

32. *Let A be an invertible* $n \times n$ *matrix. Show that* ${}^t(A^{-1}) = ({}^tA)^{-1}$.

SOLUTION. The formula proved in Exercise 9 implies

$${}^tA\,{}^t(A^{-1}) = {}^t(A^{-1}A) = {}^tI = I \quad \text{and} \quad {}^t(A^{-1})\,{}^tA = {}^t(AA^{-1}) = {}^tI = I,$$

so ${}^t(A^{-1}) = ({}^tA)^{-1}$.

33. *Let A be a complex matrix,* $A = (a_{ij})$, *and let* $\overline{A} = (\overline{a}_{ij})$, *where the bar means complex conjugate. Show that* ${}^t(\overline{A}) = \overline{{}^tA}$.

SOLUTION. The rs-entry of ${}^t(\overline{A})$ is the sr-entry of $\overline{A}$, namely, $\overline{a}_{sr}$. The rs-entry of ${}^tA$ is a_{sr}, so the rs-entry of $\overline{{}^tA}$ is $\overline{a}_{sr}$, thus ${}^t(\overline{A}) = \overline{{}^tA}$.

34. *Let A be a diagonal matrix with diagonal elements* $a_1, \dots, a_n$. *If* $a_i \neq 0$ *for all i, show that A is invertible. What is its inverse?*

SOLUTION. A brute force computation shows that

$$A^{-1} = \begin{pmatrix} a_1^{-1} & 0 & \dots & 0 \\ 0 & a_2^{-1} & \dots & 0 \\ \vdots & \vdots & & \vdots \\ 0 & 0 & \dots & a_n^{-1} \end{pmatrix}.$$

35. *Let A be a strictly upper triangular matrix, i.e. a square matrix (a_{ij}) having all its components below and on the diagonal equal to 0. Prove that if A has size $(n+1)\times(n+1)$ then $A^n = 0$. (If you wish, you may do it only in case $n = 2, 3$ and 4. The general case can be done by induction.)*

SOLUTION. The main step of the proof by induction is to show that

$$A \cdot \begin{pmatrix} 0 & \dots & 0 & b_{1k} & b_{1(k+1)} & \dots & b_{1n} \\ \vdots & & & 0 & b_{2(k+1)} & \dots & b_{2n} \\ \vdots & & & & \ddots & & \vdots \\ & & & & & & b_{(n-k+1)n} \\ & & & & & & 0 \\ & & & & & & \vdots \\ 0 & & \dots & & & & 0 \end{pmatrix} = \begin{pmatrix} 0 & \dots & 0 & 0 & c_{1(k+1)} & \dots & c_{1n} \\ \vdots & & & 0 & 0 & c_{2(k+2)} \dots & c_{2n} \\ \vdots & & & & & \ddots & \vdots \\ & & & & & & c_{(n-k)n} \\ & & & & & & 0 \\ & & & & & & 0 \\ & & & & & & \vdots \\ 0 & & \dots & & & & 0 \end{pmatrix}$$

To prove this result one simply does the computation. Let B be the second matrix on the left and C the product AB. Since $B^j = O$ for $1 \le j \le k-1$, we see that the left rectangle of size $n \times (k-1)$ of C has zeros everywhere. Note that $A_1 \cdot B^k = A_2 \cdot B^{k+1} = \dots = A_{j+1} \cdot B^{k+j} = \dots = A_{n-k+1} \cdot B^n = 0$, because the first $j+1$-entries of A_{j+1} are zero and the last $n-(j+1)$-entries of B^{k+j} are zero, so we get an additional subdiagonal of C that has all entries equal to zero. This subdiagonal is the diagonal of terms $c_{1k}, c_{2(k+1)}, \ldots, c_{n-k+1n}$. Clearly, any entry below this subdiagonal is zero.

$$\begin{pmatrix} 0 & a_{12} & a_{13} & \cdots & a_{1n} \\ 0 & 0 & a_{23} & & a_{2n} \\ \vdots & & & a_{(j+1)(j+2)} & \vdots \\ & & & \ddots & a_{(n-1)n} \\ 0 & & \cdots & & 0 \end{pmatrix}$$

$$\begin{pmatrix} 0 & \cdots & 0 & b_{1k} & b_{1(k+j)} & \cdots & b_{1n} \\ & & & 0 & b_{(j+1)(k+j)} & & \vdots \\ \vdots & & & & \ddots & & \\ & & & & & & b_{(n-k+1)n} \\ & & & & & & 0 \\ & & & & & & \vdots \\ 0 & & & \cdots & & & 0 \end{pmatrix}$$

Thus AB has the desired form, and the induction shows that $A^n = O$.

36. *Let A be a triangular matrix with components 1 on the diagonal. Let $N = A - I_n$. Show that $N^{n+1} = 0$. Note that $A = I + N$. Show that A is invertible, and that its inverse is $(I+N)^{-1} = I - N + N^2 - \ldots + (-1)^n N^n$.*

SOLUTION. Exercise 35 implies that $N^{n+1} = O$. The formula

$$(1-q)(1+q+q^2+\ldots+q^n) = 1 - q^{n+1},$$

proves that the inverse of A is $I - N + \ldots + (-1)^n N^n$. See Exercise 18.

37. *If N is a square matrix such that $N^{r+1} = 0$ for some positive integer r, show that $I - N$ is invertible and that its inverse is $I + N + \ldots + N^r$.*

SOLUTION. Again, distributing shows that

$$(I - N)(I + N + \ldots + N^n) = (I + N + \ldots + N^n)(I - N) = I - N^{n+1} = I.$$

38. *Let A be a triangular matrix:*

$$A = \begin{pmatrix} a_{11} & a_{12} & \ldots & a_{1n} \\ 0 & a_{22} & & a_{2n} \\ \vdots & \vdots & & \vdots \\ 0 & 0 & \ldots & a_{nn} \end{pmatrix}.$$

Assume that no diagonal element is 0, and let

$$B = \begin{pmatrix} a_{11}^{-1} & 0 & \ldots & 0 \\ 0 & a_{22}^{-1} & & 0 \\ \vdots & \vdots & & \vdots \\ 0 & 0 & \ldots & a_{nn}^{-1} \end{pmatrix}.$$

Show that BA ad AB are triangular matrices with components 1 on the diagonal.

SOLUTION. We compute the jth row of the product BA. We know that

$$B_j = \begin{pmatrix} 0 & \ldots & 0 & a_{jj}^{-1} & 0 & \ldots & 0 \end{pmatrix};$$

so if $k < j$, we have $B_j \cdot A^k = 0$ because the last $n - k$ are 0. Clearly $B_j \cdot A^j = 1$, so the matrix BA is triangular with diagonal entries equal to 1. A similar argument shows that AB is triangular with all diagonal entries equal to 1.

39. *A square matrix A is said to be nilpotent if $A^r = O$ for some integer $r \geq 1$. Let A, B be nilpotent matrices, of the same size, and assume $AB = BA$. Show that AB and $A + B$ are nilpotent.*

SOLUTION. Since $AB = BA$, we can manipulate the matrices A and B like numbers; hence

$$(AB)^r = A^r B^r = O,$$

and we verify at once that the binomial expansion holds, namely,

$$(A+B)^n = \sum_{k=0}^{n} \binom{n}{k} A^k B^{n-k},$$

where $\binom{n}{k} = \frac{n!}{k!(n-k)!}$. If $r \le k \le 2r$, then $A^k = O$, and if $0 \le k \le r$, then $2r - k \ge r$; so $B^{2r-k} = O$. Therefore,

$$(A+B)^{2r} = O.$$

CHAPTER III

Linear Mappings

III, §1 Mappings

1. *In Example 3, give Df as a function of x when f is the function:*
(a) $f(x)=\sin x$ *(b)* $f(x)=e^x$ *(c)* $f(x)=\log x$

SOLUTION. (a) $(Df)(x)=\cos x$ (b) $(Df)(x)=e^x$ (c) $(Df)(x)=1/x$.

2. *Prove the statement about translations in Example 13.*

SOLUTION. We have $T_{u_1+u_2}(v)=v+u_1+u_2=(v+u_2)+u_1=T_{u_1}T_{u_2}(v)$. For the second statement, note that

$$T_uT_{-u}(v)=(v-u)+u=\mathrm{id}_V(v)=(v+u)-u=T_{-u}T_u(v).$$

3. *In Example 5, give* $L(X)$ *where X is the vector:*
(a) $(1,2,-3)$ *(b)* $(-1,5,0)$ *(c)* $(2,1,1)$

SOLUTION. (a) $L(X)=11$ (b) $L(X)=13$ (c) $L(X)=6$.

4. *Let* $F\colon \mathbf{R}^2\to\mathbf{R}^2$ *be the mapping such that* $F(t)=(e^t,t)$. *What is* $F(1)$, $F(0)$, $F(-1)$?

SOLUTION. $F(1)=(e,1)$, $F(0)=(1,0)$, and $F(-1)=(e^{-1},-1)=(\frac{1}{e},-1)$.

5. *Let* $G\colon \mathbf{R}^2\to\mathbf{R}^2$ *be the mapping such that* $G(t)=(t,2t)$. *Let F be as in Exercise 4. What is* $(F+G)(1)$, $(F+G)(2)$, $(F+G)(0)$?

SOLUTION. $(F+G)(1)=(e+1,3)$, $(F+G)(2)=(e^2+2,6)$, and $(F+G)(0)=(1,0)$.

6. *Let F be as in Exercise 4. What is* $(2F)(0)$, $(\pi F)(1)$*?*

SOLUTION. $(2F)(0) = (2,0)$ and $(\pi F)(1) = (\pi e, \pi)$.

7. *Let* $A = (1,1,-1,3)$. *Let* $F: \mathbf{R}^4 \to \mathbf{R}^4$ *be the mapping such that for any vector* $X = (x_1, x_2, x_3, x_4)$ *we have* $F(X) = X \cdot A + 2$. *What is the value of* $F(X)$ *when (a)* $X = (1,1,0,-1)$ *and (b)* $X = (2,3,-1,1)$*?*

SOLUTION. (a) $F(X) = 1$ (b) $F(X) = 11$.

In Exercises 8 through 12, refer to Example 6. In each case, to prove that the image is equal to a certain set S, you must prove that the image is contained in S, and also that every element of S is in the image.

8. *Let* $F: \mathbf{R}^2 \to \mathbf{R}^2$ *be the mapping defined by* $F(x,y) = (2x, 3y)$. *Describe the image of the points lying on the circle* $x^2 + y^2 = 1$.

SOLUTION. The image of F is the ellipse whose equation is $\frac{u^2}{4} + \frac{v^2}{9} = 1$.

Indeed, if $u = 2x$, and $v = 3x$, and $x^2 + y^2 = 1$, then $\frac{u^2}{4} + \frac{v^2}{9} = 1$.

Conversely, if $\frac{u^2}{4} + \frac{v^2}{9} = 1$, and if we let $x = u/2$ and $y = u/3$, then $x^2 + y^2 = 1$ and $F(x,y) = (u,v)$.

9. *Let* $F: \mathbf{R}^2 \to \mathbf{R}^2$ *be the mapping defined by* $F(x,y) = (xy, y)$. *Describe the image under F of the straight line* $x = 2$.

SOLUTION. The image of F is the line whose equation is $y = 2x$.
Indeed, if $(2,y)$ belongs to the line $x = 2$, then $F(2,y) = (2y, y)$ and clearly $(2y, y)$ belongs to the line $y = 2x$. Conversely, suppose that $v = 2u$; then $F(2, v/2) = (v, v/2) = (v, u)$.

10. *Let F be the mapping defined by* $F(x,y) = (e^x \cos y, e^x \sin y)$. *Describe the image under F of the line* $x = 1$. *Describe more generally the image under F of a line* $x = c$, *where c is a constant.*

SOLUTION. The image of F is the circle centered at $(0,0)$ with radius e^c.
Indeed, if (c,y) belongs to the line $x = c$, then $F(c,y) = e^c(\cos y, \sin y)$.

Conversely, suppose that (u,v) belongs to the circle centered at $(0,0)$ with radius e^c, then there exists a number y such that

$$(u,v) = e^c(\cos y, \sin y).$$

Then $F(c,y) = (u,v)$.

11. *Let F be the mapping defined by* $F(t,u) = (\cos t, \sin t, u)$. *Describe geometrically the image of the* (t,u)*-plane under F.*

SOLUTION. The image of F is the cylinder in $\mathbf{R}^3$ of radius 1 with the z-axis as its major axis.

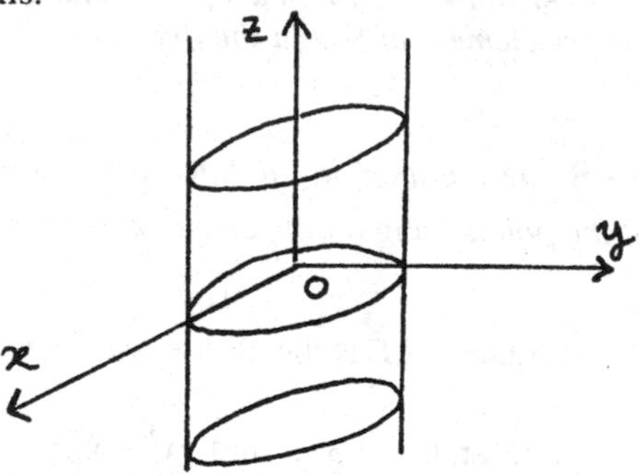

Indeed, if $F(t,u) = (a,b,c)$, then we have $a^2 + b^2 = 1$; so (a,b,c) belongs to the circle of radius one centered at the point $(0,0,c)$ and which is inscribed in the plane $z = c$.

Conversely, suppose that (a,b,c) belongs to the cylinder. Then $a^2 + b^2 = 1$, so there exists a number t such that $a = \cos t$ and $b = \sin t$; then we have $F(t,c) = (a,b,c)$.

12. *Let F be the mapping defined by* $F(x,y) = (x/3, y/4)$. *What is the image under F of the ellipse* $x^2/9 + x^2/16 = 1$?

SOLUTION. The image under F of the ellipse is the unit circle.
Indeed, suppose that $u = x/3$ and $v = y/4$; then, since (x,y) belongs to the ellipse, we see at once that $u^2 + v^2 = 1$.

Conversely, suppose that (u,v) belongs to the unit circle. Then we have $F(3u,4v) = (u,v)$ and

$$\frac{(3u)^2}{9} + \frac{(4v)^2}{16} = u^2 + v^2 = 1.$$

III, §2 Linear Mappings

1. *Determine which of the following mappings F are linear.*

(a) $F: \mathbf{R}^3 \to \mathbf{R}^2$ *defined by* $F(x, y, z) = (x, z)$

(b) $F: \mathbf{R}^4 \to \mathbf{R}^4$ *defined by* $F(X) = -X$

(c) $F: \mathbf{R}^3 \to \mathbf{R}^3$ *defined by* $F(X) = X + (0, -1, 0)$

(d) $F: \mathbf{R}^2 \to \mathbf{R}^2$ *defined by* $F(x, y) = (2x + y, y)$

(e) $F: \mathbf{R}^2 \to \mathbf{R}^2$ *defined by* $F(x, y) = (2x, y - x)$

(f) $F: \mathbf{R}^2 \to \mathbf{R}^2$ *defined by* $F(x, y) = (y, x)$

(g) $F: \mathbf{R}^2 \to \mathbf{R}$ *defined by* $F(x, y) = xy$

(h) Let U be an open set of $\mathbf{R}^3$ *and let V be the vector space of differentiable functions on U. Let V′ be the vector space of vector fields on U. Then* grad: $V \to V'$ *is a mapping. Is it linear? (For this part (h) we assume you know some calculus.)*

SOLUTION. Only the maps defined in (a), (b), (d), (e), (f), and (h) are linear. For (h), note that if f is a differentiable function on U, then we define

$$\operatorname{grad} f = \left(\frac{\partial f}{\partial x}, \frac{\partial f}{\partial y}, \frac{\partial f}{\partial z} \right).$$

The linearity of the derivative implies the linearity of grad.

2. *Let* $T: V \to W$ *be a linear map from one vector space into another. Show that* $T(O) = O$.

SOLUTION. In any vector space we have $0 \cdot v = O$, so

$$T(O) = T(0 \cdot O) = 0T(O) = O.$$

3. *Let* $T: V \to W$ *be a linear map. Let u, v be elements of V, and let* $Tu = w$. *If* $Tv = O$, *show that* $T(u + v)$ *is also equal to w.*

SOLUTION. We simply have $T(u + v) = T(u) + T(v) = w + O = w$.

4. *Let* $T: V \to W$ *be a linear map. Let U be the set of elements* $u \in V$ *such that* $T(u) = O$. *Let* $w \in W$ *and suppose there is some element* $v_0 \in V$ *such that* $T(v_0) = w$. *Show that the set of elements* $v \in V$ *satisfying* $T(v) = w$ *is precisely* $v_0 + U$.

SOLUTION. If $x \in v_0 + U$, then there exists a vector $u \in U$ such that $x = v_0 + u$, so

$$T(x) = T(v_0) + T(u) = w.$$

Conversely, suppose that $T(y) = w$. Then

$$O = w - w = T(y) - T(v_0) = T(y - v_0);$$

so $y - v_0 \in U$ thus $y \in v_0 + U$.

5. *Let $T: V \to W$ be a linear map. Let v be an element of V. Show that $T(-v) = -T(v)$.*

SOLUTION. We have $T(-v) = T((-1)v) = (-1)T(v) = -T(v)$.

6. *Let V be a vector space, and $f: V \to \mathbf{R}$, $g: V \to \mathbf{R}$ two linear mappings. Let $F: V \to \mathbf{R}^2$ be the mapping defined by $F(v) = (f(v), g(v))$. Show that F is linear. Generalize.*

SOLUTION. We prove the general result:

Theorem. Let p be a positive integer, and let $V, V^1, \ldots, V^p$ be vector spaces over a field K. For each $1 \leq j \leq p$, let $f_j: V \to V^j$ be a linear map. Then the map $F: V \to V^1 \times V^2 \times \cdots \times V^p$ defined by

$$F(v) = (f_1(v), \ldots, f_p(v))$$

is linear.

Proof. We have

$$\begin{aligned} F(v_1 + v_2) &= (f_1(v_1 + v_2), \ldots, f_p(v_1 + v_2)) = (f_1(v_1) + f_1(v_2), \ldots, f_p(v_1) + f_p(v_2)) \\ &= (f_1(v_1), \ldots, f_p(v_1)) + (f_1(v_2), \ldots, f_p(v_2)) = F(v_1) + F(v_2). \end{aligned}$$

and

$$F(cv) = (f_1(cv), \ldots, f_p(cv)) = (cf_1(v), \ldots, cf_p(v)) = c(f_1(v), \ldots, f_p(v)) = cF(v)$$

so F is linear, thereby proving the theorem.

7. *Let V, W be two vector spaces, and let $F: V \to W$ be a linear map. Let U be the subset of V consisting of all elements v such that $F(v) = O$. Prove that U is a subspace of V.*

SOLUTION. Exercise 2 implies that $O \in U$. If $v_1, v_2 \in U$, then

$$T(v_1 + v_2) = T(v_1) + T(v_2) = O,$$

so $v_1 + v_2 \in U$. If c is a number and $v \in U$, then $T(cv) = cT(v) = O$, and $cv \in U$ whence U is a subspace of V.

8. *Which of the mappings in Exercises 4, 7, 8 and 9 of §1 are linear?*

SOLUTION. The map of Exercise 8 is linear, and all the other maps are not linear.

9. *Let V be a vector space over **R**, and let $v, w \in V$. The line passing through v and parallel to w is defined to be the set of all elements $v + tw$ with $t \in \mathbf{R}$. The line segment between v and $v + w$ is defined to be the set of all elements $v + tw$ with $0 \le t \le 1$. Let $L: V \to U$ be a linear map. Show that the image under L of a line segment in V is a line segment in U. Between what points?. Show that the image of a line under L is either a line or a point.*

SOLUTION. The image of the line segment between v and $v + w$ is the line segment between $L(v)$ and $L(v) + L(w)$ because

$$L(v + tw) = L(v) + tL(w).$$

From this expression we also see that the image of a line passing through v and parallel to w is a point when $L(w) = O$ and a line when $L(w) \ne O$.

Let V be a vector space, and let v_1, v_2 be two elements of V which are linearly independent. The set of elements of V which can be written in the form $t_1v_1 + t_2v_2$ with numbers t_1, t_2 satisfying $0 \le t_1 \le 1$ and $0 \le t_2 \le 1$ is called the parallelogram spanned by v_1, v_2.

10. *Let V and W be vector spaces, and let $F: V \to W$ be a linear map. Let v_1, v_2 be linearly independent elements of V, and assume that $F(v_1)$, $F(v_2)$ are linearly independent. Show that the image under F of the parallelogram spanned by v_1 and v_2 is the parallelogram spanned by $F(v_1)$, $F(v_2)$.*

SOLUTION. The result is a consequence of the linearity of F, namely,

$$F(t_1v_1 + t_2v_2) = t_1F(v_1) + t_2F(v_2).$$

11. *Let F be a linear map of $\mathbf{R}^2$ into itself such that $F(E_1) = (1,1)$ and $F(E_2) = (-1,2)$. Let S be the square whose corners are at $(0,0)$, $(1,0)$, $(1,1)$ and $(0,1)$. Show that the image of this square under F is a parallelogram.*

SOLUTION. The square is the set of elements of $\mathbf{R}^2$ that can be written in the form $t_1E_1 + t_2E_2$, where $0 \le t_1 \le 1$ and $0 \le t_2 \le 1$. Then we have

$$F(t_1E_1 + t_2E_2) = t_1(1,1) + t_2(-1,2).$$

The vectors $(1,1)$ and $(-1,2)$ are linearly independent because $1 \times 2 + 1 \times 1 \ne 0$ (see Exercise 4 in §2 of Chapter I), so the image of the square under F is the parallelogram spanned by $(1,1)$ and $(-1,2)$.

12. *Let A, B be two non-zero vectors in the plane such that there is no constant $c \ne 0$ such that $B = cA$. Let T be a linear mapping of the plane into itself such that $T(E_1) = A$ and $T(E_2) = B$. Describe the image under T of the rectangle whose corners are $(0,1)$, $(3,0)$, $(0,0)$, $(3,1)$.*

SOLUTION. Exercise 3 in §4 of Chapter I implies that A and B are linearly independent. The rectangle is the set of vectors in $\mathbf{R}^2$ that can be written in the form $t_1(3E_1) + t_2E_2$, where $0 \le t_1 \le 1$ and $0 \le t_2 \le 1$. We have

$$T(t_1(3E_1) + t_2E_2) = t_1(3A) + t_2B,$$

so the image of the rectangle under T is the parallelogram spanned by $3A$ and B.

13. *Let A, B be two non-zero vectors in the plane such that there is no constant $c \ne 0$ such that $B = cA$. Describe geometrically the set of points $tA + uB$ for values of t and u such that $0 \le t \le 5$ and $0 \le u \le 2$.*

SOLUTION. Let $t_1 = t/5$ and $t_2 = u/2$. Then $0 \le t_1 \le 1$, $0 \le t_2 \le 1$, and

$$tA + uB = t_1(5A) + t_2(2B).$$

Exercise 3 in §4 of Chapter I implies that A and B are linearly independent, so the set in question is simply the parallelogram spanned by $5A$ and $2B$.

14. *Let $T_u : V \to V$ be the translation by a vector u. For which vector u is T_u a linear map? Proof?*

SOLUTION. The translation by a vector u is a linear map if and only if $u = O$. Indeed, suppose that T_u is a linear map. Then $T_u(O) = O$, and so $u = O$. Conversely, if $u = O$, then T_u is the identity, which is linear.

15. *Let V, W be two vector spaces and, $F\colon V \to W$ a linear map. Let $w_1, \dots, w_n$ be elements of W which are linearly independent, and let $v_1, \dots, v_n$ be elements of V such that $F(v_i) = w_i$ for $i = 1, \dots, n$. Show that $v_1, \dots, v_n$ are linearly independent.*

SOLUTION. Suppose that $a_1 v_1 + \dots + a_n v_n = O$. Then

$$O = T(a_1 v_1 + \dots + a_n v_n) = a_1 w_1 + \dots + a_n w_n,$$

therefore, $a_1 = \dots = a_n = 0$.

16. *Let V be a vector space and $F\colon V \to \mathbf{R}$ a linear map. Let W be the subset of V consisting of all elements v such that $F(v) = O$. Assume that $W \neq V$ and let v_0 be an element of V which does not lie in W. Show that every element of V can be written as a sum $w + cv_0$ with some w in W and some number c.*

SOLUTION. Let x be an element of V and since $T(v_0) \neq 0$, we let $c = T(x)/T(v_0)$. Finally, let $w = x - cv_0$. All we have to do is check that $w \in W$. The definition of c implies

$$T(w) = T(x - cv_0) = T(x) - cT(v_0) = 0.$$

17. *In Exercise 16, show that W is a subspace of V. Let $\{v_1, \dots, v_n\}$ be a basis of W. Show that $\{v_0, v_1, \dots, v_n\}$ is a basis of V.*

SOLUTION. See Exercise 7 for a proof that W is a subspace. Exercise 16 shows that the set $\{v_0, v_1, \dots, v_n\}$ generates V. Since $\{v_1, \dots, v_n\}$ are linearly independent, and $v_0 \notin W$ and the vectors $\{v_0, v_1, \dots, v_n\}$ are linearly independent, we conclude that $\{v_0, v_1, \dots, v_n\}$ is a basis for V. One can give another proof. Consider the relation $a_0 v_0 + a_1 v_1 + \dots + a_n v_n = O$. Take its image under T to find that $a_0 = 0$. Then the fact that $\{v_1, \dots, v_n\}$ are linearly independent concludes the argument.

18. *Let $L\colon \mathbf{R}^2 \to \mathbf{R}^2$ be a linear map, having the following effect on the indicated vectors:*

(a) $L(3,1)=(1,2)$ *and* $L(-1,0)=(1,1)$
(b) $L(4,1)=(1,1)$ *and* $L(1,1)=(3,-2)$
(c) $L(1,1)=(2,1)$ *and* $L(-1,1)=(6,3)$.
In each case compute $L(1,0)$.

SOLUTION.
(a) $L(1,0)=(-1,-1)$ because $(1,0)=-(-1,0)$.

(b) $L(1,0)=(-2/3,1)$ because $(1,0)=\frac{1}{3}(4,1)-\frac{1}{3}(1,1)$.

(c) $L(1,0)=(-2,-1)$ because $(1,0)=\frac{1}{2}(1,1)-\frac{1}{2}(-1,1)$.

19. *Let L be as in (a), (b), (c), of Exercise 18. Find* $L(0,1)$.

SOLUTION.
(a) $L(0,1)=(4,5)$ because $(0,1)=(3,1)+3(-1,0)$.

(b) $L(0,1)=(\frac{11}{3},-3)$ because $(0,1)=\frac{-1}{3}(4,1)+\frac{4}{3}(1,1)$.

(c) $L(0,1)=(4,2)$ because $(0,1)=\frac{1}{2}(1,1)+\frac{1}{2}(-1,1)$.

III, §3 The Kernel and Image of a Linear Map

1. *Let A, B be two vectors in* $\mathbf{R}^2$ *forming a basis of* $\mathbf{R}^2$. *Let* $F\colon \mathbf{R}^2 \to \mathbf{R}^n$ *be a linear map. Show that either* $F(A)$, $F(B)$ *are linearly independent, or the image of F has dimension 1, or the image of F is* $\{O\}$.

SOLUTION. We have the relation

$$2=\dim \operatorname{Im} F+\dim \operatorname{Ker} F,$$

so if $F(A)$ and $F(B)$ are linearly dependent, then $\dim \operatorname{Ker} F \geq 1$, and thus $\dim \operatorname{Im} F \leq 1$.

2. *Let A be a non-zero vector in* $\mathbf{R}^2$. *Let* $F\colon \mathbf{R}^2 \to W$ *be a linear map such that* $F(A)=O$. *Show that the image of F is either a straight line or* $\{O\}$.

SOLUTION. Since $F(tA)=tF(A)=O$, we have $\dim \operatorname{Ker} F \geq 1$. But $\dim \mathbf{R}^2=2$, so $\dim \operatorname{Im} F$ is 0 or 1 whence the image of F is either $\{O\}$ or a straight line.

3. *Determine the dimension of the subspace of* $\mathbf{R}^4$ *consisting of all* $X \in \mathbf{R}^4$ *such that* $x_1 + 2x_2 = 0$ *and* $x_3 - 15x_4 = 0$.

SOLUTION. Let W be the subspace in question. Then $\dim W = 2$. One can see that $(-2, 1, 0, 0)$ and $(0, 0, 15, 1)$ form a basis for W. Or we may consider the linear map $L\colon \mathbf{R}^4 \to \mathbf{R}^2$ defined by

$$L(x_1, x_2, x_3, x_4) = (x_1 + 2x_2, x_3 - 15x_4).$$

Then $\operatorname{Ker} L = W$ and $\operatorname{Im} L = \mathbf{R}^2$.

4. *Let* $L\colon V \to W$ *be a linear map. Let* w *be an element of* W. *Let* v_0 *be an element of* V *such that* $L(v_0) = w$. *Show that any solution of the equation* $L(X) = w$ *is of type* $v_0 + u$, *were* u *is an element of the kernel of* L.

SOLUTION. See Exercise 4 in §2.

5. *Let* V *be the vector space of all functions which have derivatives of all orders and let* $D\colon V \to V$ *be the derivative. What is the kernel of* D?

SOLUTION. The kernel of D is the set of all constant functions.

6. *Let* D^2 *be the second derivative (i.e. the iteration of* D *taken twice). What is the kernel of* D^2? *In general, what is the kernel of* D^n *(n-th derivative)?*

SOLUTION. By integration we see that the kernel of D^n is the set of polynomials of degree $\leq n-1$.

7. *Let* V *be again the vector space of functions which have derivatives of all orders. Let* W *be the subspace of* V *consisting of those functions* f *such that* $f'' + 4f = 0$ *and* $f(\pi) = 0$. *Determine the dimension of* W.

SOLUTION. The space W has dimension 1. We use a trick to prove the following theorem which is a special case of a more general theorem on differential equations.

Theorem. Let c be a positive number. Then $(\cos ct, \sin ct)$ is a basis for the solution space of infinitely differentiable functions of the second-order differential equation

$$f'' + c^2 f = 0. \qquad (*)$$

Proof. The functions $\cos ct$ and $\sin ct$ are solutions of $(*)$, and they are linearly independent. Let f be an infinitely differentiable function solution of $(*)$. Then differentiating the expressions

$$f(t)\cos ct - \frac{1}{c} f'(t)\sin ct \quad \text{and} \quad f(t)\sin ct + \frac{1}{c} f'(t)\cos ct,$$

and using $(*)$, one finds 0. So there exist constants a and b such that

$$\begin{cases} f(t)\cos ct - \dfrac{1}{c} f'(t)\sin ct = a \\ f(t)\sin ct + \dfrac{1}{c} f'(t)\cos ct = b \end{cases}$$

Multiplying the first equation by $\cos ct$, the second by $\sin ct$, and then adding the resulting equations, we find that $f(t) = a\cos ct + b\sin ct$; thereby concluding the proof of the theorem.

Back to the original problem we see that the condition $f(\pi) = 0$ implies that $a = 0$; so f must be of the form $b\sin 2t$, and therefore W has dimension 1.

8. *Let V be the vector space of all infinitely differentiable function. We write the functions as function of a variable t, and let $D = d/dt$. Let $a_1, \ldots, a_m$ be numbers. Let g be an element of V. Describe how the problem of finding a solution of the differential equation*

$$a_m \frac{d^m f}{dt^m} + a_{m-1} \frac{d^{m-1} f}{dt^{m-1}} + \ldots + a_0 f = g$$

can be interpreted as fitting the abstract situation described in Exercise 6.

SOLUTION. Let S be the set of solutions of the same equation but with 0 instead of g. The map

$$L\colon \mathrm{V} \to \mathrm{V}$$

$$f \mapsto a_m \frac{d^m f}{dt^m} + \ldots + a_0 f$$

is linear and $\operatorname{Ker} L = S$. So if we have one solution, say f_0, of the equation

$$a_m \frac{d^m f}{dt^m} + \ldots + a_0 f = g \quad (*)$$

then the general solution of $(*)$ can be written as $f_0 + h$, where h lies in S.

9. *Again let V be the space of all infinitely differentiable functions, and let $D: V \to V$ be the derivative.*
(a) Let $L = D - I$ where I is the identity mapping. What is the kernel of L?
(b) Same question if $L = D - aI$, where a is a number.

SOLUTION. (a) and (b) We want to solve

$$f' - af = 0. \quad (*)$$

Let f be a solution of $(*)$. Then

$$\left(\frac{f(t)}{e^{at}}\right)' = \frac{f'(t)e^{at} - af(t)e^{at}}{e^{2at}} = \frac{e^{at}\left(f'(t) - af(t)\right)}{e^{2at}} = 0.$$

So there exists a constant c such that $f(t) = ce^{at}$. Conversely, any function of this form solves $(*)$, so $\operatorname{Ker} L$ is the space generated by e^{at}.

10. *(a) What is the dimension of the subspace of K^n consisting of those vectors $A = (a_1, \dots, a_n)$ such that $a_1 + \dots + a_n = 0$?*
(b) What is the dimension of the subspace of the space of $n \times n$ matrices (a_{ij}) such that $a_{11} + \dots + a_{nn} = 0$?
[For part (b) look at the next exercise.]

SOLUTION. (a) The dimension of the space in question is $n-1$. Consider the linear map $L: K^n \to K$ defined by $L(A) = a_1 + \dots + a_n$. Then clearly $\operatorname{Im} L = K$ because, given any $x \in K$, then $L(x, 0, \dots, 0) = x$. Thus $\dim \operatorname{Ker} L = n-1$. Note that the set $\{E_j - E_n\}_{1 \le j \le n-1}$ is a basis for the space in question because $a_n = -a_1 - \dots - a_{n-1}$.

(b) The dimension of the space in question is $n^2 - 1$. Indeed, the linear map defined in Exercise 11(a) has an image equal to K because, given any $x \in K$, the matrix with all entries 0 except $a_{11} = x$ has trace equal to x. Since the space of square $n \times n$ matrices is n^2, the result follows.

11. *Let $A = (a_{ij})$ be an $n \times n$ matrix. Define the trace of A to be the sum of the diagonal elements.*
(a) Show that the trace is a linear map of the space of $n \times n$ matrices into K.
(b) If A, B are $n \times n$ matrices, show that $\operatorname{tr}(AB) = \operatorname{tr}(BA)$.
(c) If B is invertible show that $\operatorname{tr}(B^{-1}AB) = \operatorname{tr}(A)$.
(d) If A, B are $n \times n$ matrices, show that the association

$$(A, B) \mapsto \operatorname{tr}(AB) = \langle A, B\rangle$$

satisfies the three conditions of a scalar product.

(e) Prove that there are no matrices A, B such that $AB - BA = I_n$.

SOLUTION. (a) If $B = (b_{ij})$, then

$$\operatorname{tr}(A+B) = \sum_{i=1}^{n}(a_{ii} + b_{ii}) = \sum_{i=1}^{n} a_{ii} + \sum_{i=1}^{n} b_{ii} = \operatorname{tr}(A) + \operatorname{tr}(B).$$

and if c is a number,

$$\operatorname{tr}(cA) = \sum_{i=1}^{n} ca_{ii} = c\sum_{i=1}^{n} a_{ii} = c\operatorname{tr}(A).$$

(b) See Exercise 27 in §3 of Chapter II.

(c) By (b) we have

$$\operatorname{tr}(B^{-1}AB) = \operatorname{tr}(B^{-1}(AB)) = \operatorname{tr}((AB)B^{-1}) = \operatorname{tr}(ABB^{-1}) = \operatorname{tr}(A).$$

(d) The result in (c) implies **SP 1**. The properties **SP 2** and **SP 3** are verified because of (a).

(e) Suppose that there exist matrices A and B such that $AB - BA = I_n$. Then

$$\operatorname{tr}(AB - BA) = \operatorname{tr}(I_n) = n,$$

but (a) and (b) imply that

$$\operatorname{tr}(AB - BA) = \operatorname{tr}(AB) - \operatorname{tr}(BA) = 0,$$

so $0 = n$, which is a contradiction.

12. *Let S be the set of symmetric* $n \times n$ *matrices. Show that S is a vector space. What is the dimension of S? Exhibit a basis for S, when* $n = 2$ *and* $n = 3$.

SOLUTION. If $a_{ij} = a_{ji}$ and $b_{ij} = b_{ji}$, then $a_{ij} + b_{ij} = a_{ji} + b_{ji}$, so the sum of two elements of S belongs to S. Clearly, the product of an element of S by a scalar is an element of S, and the zero matrix belongs to S, so S is a vector space.

We proved in Exercise 6, §1 of Chapter II that the dimension of S is $\frac{n(n+1)}{2}$, and we also gave a basis for the general case. When $n=2$, a basis is given by the three matrices

$$\begin{pmatrix} 0 & 1 \\ 1 & 0 \end{pmatrix}, \quad \begin{pmatrix} 1 & 0 \\ 0 & 0 \end{pmatrix}, \quad \begin{pmatrix} 0 & 0 \\ 0 & 1 \end{pmatrix}.$$

When $n=3$, a basis is given by the six matrices

$$\begin{pmatrix} 0 & 1 & 0 \\ 1 & 0 & 0 \\ 0 & 0 & 0 \end{pmatrix}, \quad \begin{pmatrix} 0 & 0 & 1 \\ 0 & 0 & 0 \\ 1 & 0 & 0 \end{pmatrix}, \quad \begin{pmatrix} 0 & 0 & 0 \\ 0 & 0 & 1 \\ 0 & 1 & 0 \end{pmatrix},$$

$$\begin{pmatrix} 1 & 0 & 0 \\ 0 & 0 & 0 \\ 0 & 0 & 0 \end{pmatrix}, \quad \begin{pmatrix} 0 & 0 & 0 \\ 0 & 1 & 0 \\ 0 & 0 & 0 \end{pmatrix}, \quad \begin{pmatrix} 0 & 0 & 0 \\ 0 & 0 & 0 \\ 0 & 0 & 1 \end{pmatrix}.$$

13. *Let A be a real symmetric $n \times n$ matrix. Show that* $\operatorname{tr}(AA) \geq 0$ *and if* $A \neq O$, *then* $\operatorname{tr}(AA) > 0$.

SOLUTION. The kth diagonal entry of AA is given by the scalar product $A_k \cdot A^k$. But A is symmetric, so $A_k = {}^t A^k$. Hence $c_{kk} \geq 0$ and $c_{kk} = 0$ if and only if $A_k = O$. Therefore, $\operatorname{tr}(AA) \geq 0$ and $\operatorname{tr}(AA) > 0$ whenever $A \neq O$.

14. *An $n \times n$ matrix is called skew-symmetric if ${}^t A = -A$. Show that any $n \times n$ matrix A can be written as a sum $A = B + C$, where B is symmetric and C is skew-symmetric. [Hint: Let $B = (A + {}^t A)/2$.] Show that if $A = B_1 + C_1$ where B_1 is symmetric and C_1 is skew-symmetric, then $B = B_1$ and $C = C_1$.*

SOLUTION. Let $2B = A + {}^t A$ and $2C = A - {}^t A$. The matrix B is symmetric because

$${}^t(2B) = {}^t(A + {}^t A) = {}^t A + {}^t({}^t A) = {}^t A + A = 2B,$$

and C is skew-symmetric because

$${}^t(2C) = {}^t(A - {}^t A) = {}^t A - {}^t({}^t A) = -(A - {}^t A) = -2C.$$

Note that $2A = (A+{}^tA)+(A-{}^tA) = 2B+2C$, so $A = B+C$. Moreover, if $A = B_1 + C_1$, where B_1 is symmetric and C_1 is skew-symmetric, then $B_1 - B$ is symmetric, $C - C_1$ is skew-symmetric, and $B_1 - B = C - C_1$, so

$$(B_1 - B) = {}^t(B_1 - B) = {}^t(C - C_1) = C_1 - C = B - B_1.$$

Thus $2(B_1 - B) = O$. So we find that $B = B_1$ and $C = C_1$. Therefore we have the general result:

The space of $n \times n$ matrices is the direct sum of the space of symmetric $n \times n$ matrices and the space of $n \times n$ skew-symmetric matrices.

15. *Let M be the space of all $n \times n$ matrices. Let $P\colon M \to M$ be the map such that*

$$P(A) = \frac{A+{}^tA}{2}.$$

(a) Show that P is linear.
(b) Show that the kernel of P consists of the space of skew-symmetric matrices.
(c) What is the dimension of the kernel of P?

SOLUTION. (a) The map *P* is linear because

$$P(A+B) = \frac{(A+B)+{}^t(A+B)}{2} = \frac{A+{}^tA + B+{}^tB}{2} = P(A)+P(B),$$

and if *c* is a scalar,

$$P(cA) = \frac{cA+{}^t(cA)}{2} = c\frac{A+{}^tA}{2} = cP(A).$$

(b) The equation $P(A) = O$ is equivalent to $A+{}^tA = O$, which is equivalent to ${}^tA = -A$.

(c) If $\mathrm{Sym}_n(K)$ is the set of symmetric $n \times n$ matrices and $\mathrm{Sk}_n(K)$ is the set of skew symmetric matrices, then Exercise 14 implies

$$\mathrm{Mat}_{n\times n}(K) = \mathrm{Sym}_n(K) \oplus \mathrm{Sk}_n(K),$$

thus $\dim \mathrm{Mat}_{n\times n}(K) = \dim \mathrm{Sym}_n(K) + \dim \mathrm{Sk}_n(K)$. Exercise 6 in §1 of Chapter II implies that

$$\dim \operatorname{Ker} P = n^2 - \frac{n(n+1)}{2} = \frac{n(n-1)}{2}.$$

16. *Let M be the space of all $n \times n$ matrices. Let $F: M \to M$ be the map such that*

$$F(A) = \frac{A - {}^tA}{2}.$$

(a) Show that F is linear.
(b) Describe the kernel of Fm and determine its dimension.

SOLUTION. (a) Argue as in Exercise 15 to show that F is linear.

(b) The kernel of F is the set of symmetric matrices that has dimension $\frac{n(n+1)}{2}$.

17. *(a) Let U, W be vector spaces. We let $U \times W$ be the set of all pairs (u, w) with $u \in U$ and $w \in W$. If (u_1, w_1), (u_2, w_2) are such pairs, define their sum*

$$(u_1, w_1) + (u_2, w_2) = (u_1 + u_2, w_1 + w_2).$$

If c is a number, define $c(u, w) = (cu, cw)$. Show that $U \times W$ is a vector space with these definitions. What is the zero element?
(b) If U has dimension n and W has dimension m, what is the dimension of $U \times W$? Exhibit a basis of $U \times W$ in terms of a basis for U and a basis for W.
(c) If U is a subspace of a vector space V, show that the subset of $V \times V$ consisting of all elements (u, u) with $u \in U$ is a subspace.

SOLUTION. (a) The set $U \times W$ is a vector space because U and W are vector spaces, and its zero element is $O_{U \times W} = (O_U, O_W)$, which we write as (O, O).

(b) We have $\dim U \times V = \dim U + \dim V = n + m$. If $\{u_1, \ldots, u_n\}$ is a basis for U and $\{w_1, \ldots, w_m\}$ is a basis for W, then

$$\{(u_1, O), \ldots, (u_n, O), (O, w_1), \ldots, (O, w_m)\}$$

is a basis for $U \times W$.

(c) Let U' be the set in question. Then $(O,O) \in U'$ and if (u_1,u_1) and (u_2,u_2) belong to U', then $(u_1+u_2,u_1+u_2) \in U'$ and $(cu_1,cu_1) \in U'$ because U is a vector space, whence U' is subspace of $V \times V$.

18. *(To be done after you have done Exercise 17.) Let U, W be subspaces of a vector space V. Show that*

$$\dim U + \dim W = \dim (U+W) + \dim (U \cap W).$$

[Hint: Show that the map $L\colon U \times W \to V$ given by $L(u,w) = u - w$ is a linear map. What is its image? What is its kernel?]

SOLUTION. We have

$$L(u_1+u_2, w_1+w_2) = u_1+u_2-w_1-w_2 = L(u_1,w_1)+L(u_2,w_2)$$

and

$$L(cu, cw) = cu - cw = cL(u,w),$$

so the map L is linear. We investigate the image and kernel of L:

Image of L. Clearly, $\operatorname{Im} L \subset U+W$ and, conversely, given $u \in U$ and $u \in W$, we see that $L(u,-w) = u+w$; so

$$\operatorname{Im} L = U + W.$$

Kernel of L. $L(u,w) = O$ if and only if $u = w$, so

$$\operatorname{Ker} L = U \cap W.$$

Therefore, $\dim (U \times W) = \dim (U+W) + \dim (U \cap W)$. Conclude the argument using (b) of Exercise 17.

III, §4 Composition and Inverse of Linear Mappings

1. *Let $L\colon \mathbf{R}^2 \to \mathbf{R}^2$ be a linear map such that $L \neq O$ but $L^2 = LL = 0$. Show that there exists a basis $\{A, B\}$ of $\mathbf{R}^2$ such that $L(A) = B$ and $L(B) = O$.*

SOLUTION. By assumption, there exists non-zero vectors A and B such that $L(A)=B$. Then $L(B)=L^2(A)=O$ and if $aA+bB=O$, then

$$O=L(aA+bB)=aL(A),$$

so $a=b=0$.

2. *Let* $\dim V>\dim W$. *Let* $L\colon V\to W$ *be a linear map. Show that the kernel of L is not* $\{O\}$.

SOLUTION. If $\dim \operatorname{Ker} L=0$, then we have $\dim V=\dim \operatorname{Im} L$. But $\dim \operatorname{Im} L\le \dim W$, which implies that $\dim V\le \dim \operatorname{Im} W$, so we have a contradiction.

3. *Finish the proof of Theorem 4.3.*

SOLUTION. Let $G(v)=u$, then $F(cu)=cF(u)=cv$. Compose with G and conclude.

4. *Let* $\dim V=\dim W$. *Let* $L\colon V\to W$ *be a linear map whose kernel is* $\{O\}$. *Show that L has an inverse linear map.*

SOLUTION. Theorem 3.2 implies that $\dim \operatorname{Im} L=\dim W$, so L is surjective. Conclude.

5. *Let F, G be invertible linear maps of a vector space V onto itself. Show that* $(FG)^{-1}=G^{-1}F^{-1}$.

SOLUTION. We have $(FG)(G^{-1}F^{-1})=F(GG^{-1})F^{-1}=FF^{-1}=I$, and, similarly, we see that $(G^{-1}F^{-1})(FG)=I$.

6. *Let* $L\colon \mathbf{R}^2\to\mathbf{R}^2$ *be the linear map defined by* $L(x,y)=(x+y,x-y)$. *Show that L is invertible.*

SOLUTION. If $L(x,y)=O$, then $x=-y$ and $x=y$; so $\operatorname{Ker} L=\{O\}$.

7. *Let* $L\colon \mathbf{R}^2\to\mathbf{R}^2$ *be the linear map defined by* $L(x,y)=(2x+y,3x-5y)$. *Show that L is invertible.*

SOLUTION. Verify that $\operatorname{Ker} L=\{O\}$.

8. *Let* $L\colon \mathbf{R}^3\to\mathbf{R}^3$ *be the linear maps as indicated. Show that L is invertible in each case.*

(a) $L(x, y, z) = (x - 2y, x + z, x + y + 2z)$
(b) $L(x, y, z) = (2x - y + z, x + y, 3x + y + z)$

SOLUTION. (a) Since $\operatorname{Ker} L = \{O\}$, L is invertible.

(b) Since $\operatorname{Ker} L = \{O\}$, L is invertible.

9. *(a) Let $L: V \to V$ be a linear mapping such that $L^2 = O$. Show that $I - L$ is invertible. (I is the identity mapping on V.)*
(b) Let $L: V \to V$ be a linear map such that $L^2 + 2L + I = O$. Show that L is invertible.
(c) Let $L: V \to V$ be a linear map such that $L^3 = O$. Show that $I - L$ is invertible.

SOLUTION. (a) The inverse of L is $I + L$.

(b) The inverse of L is $-L - 2$.

(c) The inverse of L is $I + L + L^2$.

10. *Let V be a vector space. Let $P: V \to V$ be a linear map such that $P^2 = P$. Show that*

$$V = \operatorname{Ker} P + \operatorname{Im} P \quad \textit{and} \quad \operatorname{Ker} P \cap \operatorname{Im} P = \{O\}$$

in other words, V is the direct sum of Ker P *and* Im P. *[Hint: To show V is the sum, write an element of V in the form $v = v - Pv + Pv$.]*

SOLUTION. Since $P(v - P(v)) = P(v) - P^2(v) = O$, we see that $V = \operatorname{Ker} P + \operatorname{Im} P$. As for the intersection, note that if w lies in the set $\operatorname{Ker} P \cap \operatorname{Im} P$, then there exists a vector v such that $P(v) = w$, so that $P(v) = P(w)$. But since $P(w) = O$, we conclude that $w = P(v) = O$. Hence

$$V = \operatorname{Ker} L \oplus \operatorname{Im} L.$$

11. *Let V be a vector space and P, Q be linear maps of V into itself. Assume that they satisfy the following conditions:*
(a) $P + Q = I$ (identity mapping).
(b) $PQ = QP = O$.
(c) $P^2 = P$ and $Q^2 = Q$.
Show that V is the direct sum of Im P *and* Im Q.

SOLUTION. In Exercise 12 we prove that $\operatorname{Im} P = \operatorname{Ker} Q$, and in Exercise 10 we prove that $V = \operatorname{Ker} Q \oplus \operatorname{Im} Q$, so the result drops out.

12. *Notations being as in Exercise 11, show that the image of P is equal to the kernel of Q. [Prove the two statements:*

Image of P is contained in kernel of Q.
Kernel of Q contained in image of P.]

SOLUTION. If $v \in \operatorname{Im} P$, then there exists w such that $P(w) = v$. Then $QP(w) = Q(v)$, so $v \in \operatorname{Ker} Q$.

Conversely, suppose that $v \in \operatorname{Ker} Q$. Then $v = P(v) + Q(v) = P(v)$, so $v \in \operatorname{Im} P$.

13. *Let $T\colon V \to V$ be a linear map such that $T^2 = I$. Let*

$$P = \tfrac{1}{2}(I + T) \quad \text{and} \quad Q = \tfrac{1}{2}(I - T).$$

Prove:

$$P + Q = I, \quad P^2 = P, \quad Q^2 = Q, \quad PQ = QP = O.$$

SOLUTION. We have

$$P + Q = I + \tfrac{1}{2}T - \tfrac{1}{2}T = I$$
$$P^2 = \tfrac{1}{4}(I + T)^2 = \tfrac{1}{4}(I + 2T + I) = P$$
$$PQ = \tfrac{1}{4}\left(I - T^2\right) = O.$$

Similarly, $Q^2 = Q$ and $QP = O$.

14. *Let $F\colon V \to W$ and $G\colon W \to U$ be isomorphisms of vector spaces over K. Show that GF is invertible, and that $(GF)^{-1} = F^{-1}G^{-1}$.*

SOLUTION. See Exercise 5.

15. *Let $F\colon V \to W$ and $G\colon W \to U$ be isomorphisms of vector spaces over K. Show that $GF\colon V \to U$ is an isomorphism.*

SOLUTION. See Exercise 14.

16. *Let V, W be two vector spaces over K, of finite dimension n. Show that V and W are isomorphic.*

SOLUTION. Let $\{v_1, \ldots, v_n\}$ be a basis for V and $\{w_1, \ldots, w_n\}$ a basis for W. Then the mapping $L\colon V \to W$ defined by

$$L\left(a_1 v_1 + \ldots + a_n v_n\right) = a_1 w_1 + \ldots + a_n w_n$$

is an isomorphism.

17. *Let A be a linear map of a vector space into itself, and assume that*

$$A^2 - A + I = O$$

(where I is the identity map). Show that A^{-1} exists and is equal to $I - A$. Generalize (cf. Exercise 37 of Chapter II, §3)

SOLUTION. Replace N by A in the answer to Exercise 37 in §3 of Chapter II.

18. *Let A, B be linear maps of a vector space into itself. Assume that $AB = BA$. Show that*

$$(A+B)^2 = A^2 + 2AB + B^2$$

and

$$(A+B)(A-B) = A^2 - B^2$$

SOLUTION. We have

$$(A+B)(A+B) = A(A+B) + B(A+B) = A^2 + 2AB + B^2$$

and

$$(A+B)(A-B) = A(A-B) + B(A-B) = A^2 - B^2.$$

19. *Let A, B be linear maps of a vector space into itself. If the kernel of A is $\{O\}$ and the kernel of B is $\{O\}$, show that the kernel of AB is $\{O\}$.*

SOLUTION. See Exercise 20.

20. *More generally, let $A\colon V \to W$ and $B\colon W \to U$ be linear maps. Assume that the kernel of A is $\{O\}$ and the kernel of B is $\{O\}$, show that the kernel of BA is $\{O\}$.*

SOLUTION. Suppose that $BA(v) = O$. Then $A(v) \in \operatorname{Ker} B$, so $A(v) = O$, thus $v = O$.

21. *Let $A\colon V \to W$ and $B\colon W \to U$ be linear maps. Assume that A is surjective and that B is surjective. Show that BA is surjective.*

SOLUTION. Given $u \in U$, there exists an element $w \in W$ such that $B(w) = u$ and there exists an element $v \in V$ such that $A(v) = w$. Then $BA(v) = u$.

III, §5 Geometric Applications

1. *Show that the image under a linear map of a convex set is a convex set.*

SOLUTION. Let S be a convex set, and let L be a linear map. Let $S' = L(S)$. Suppose that $w_1, w_2 \in S'$, and let $v_1, v_2 \in S$ be such that $L(v_i) = w_i$. By definition, the line segment $tv_1 + (1-t)v_2$, $t \in [0,1]$, is contained in S, thus $tw_1 + (1-t)w_2$ belongs to S' because

$$tw_1 + (1-t)w_2 = L(tv_1 + (1-t)v_2).$$

2. *Let S_1 and S_2 be convex sets in V. Show that the intersection $S_1 \cap S_2$.*

SOLUTION. Let $v, w \in S_1 \cap S_2$. Then v and W belong to S_1, so the line segment between v and w is contained in S_1. Similarly, this line segment is contained in S_2, so the line segment between v and w is contained in $S_1 \cap S_2$.

3. *Let $L: \mathbf{R}^n \to \mathbf{R}$ be a linear map. Let S be the set of all points A in $\mathbf{R}^n$ such that $L(A) \geq 0$. Show that S is convex.*

SOLUTION. The set $\mathbf{R}_{\geq 0}$ is convex. Apply Exercise 6.

4. *Let $L: \mathbf{R}^n \to \mathbf{R}$ be a linear map and c a number. Show that the set S consisting of all points A in $\mathbf{R}^n$ such that $L(A) > c$ is convex.*

SOLUTION. The set (c, ∞) is convex. Apply Exercise 6.

5. *Let A be a non-zero vector in $\mathbf{R}^n$ and c a number. Show that the set of points X such that $X \cdot A \geq c$ is convex.*

SOLUTION. The set $[c, \infty)$ is convex and the map $L: \mathbf{R}^n \to \mathbf{R}$, $X \to X \cdot A$ is linear. Apply Exercise 6.

6. *Let $L: V \to W$ be a linear map. Let S' be a convex set in W. Let S be the set of all elements P in V such that $L(P)$ is in S'. Show that S is convex.*

SOLUTION. Let $P, Q \in S$. Then by assumption, $tL(P) + (1-t)L(Q) \in S'$ whenever $t \in [0,1]$. Hence $L(tP + (1-t)Q) \in S'$, thus $tP + (1-t)Q \in S$.

7. *Show that a parallelogram is convex.*

SOLUTION. Suppose that $P = uv + sw$ and $Q = u'v + s'w$ belong to the parallelogram spanned by v and w. Then for $t \in [0,1]$ we have

$$tP + (1-t)Q = \left(tu + (1-t)u'\right)v + \left(ts + (1-t)s'\right)w.$$

But $0 \le tu + (1-t)u' \le 1$ and $0 \le ts + (1-t)s' \le 1$, so the triangle spanned by v and w is convex. Exercise 8 then implies that any parallelogram is convex.

8. *Let S be a convex set in V and let u be an element of V. Let $T_u\colon V \to V$ be the translation by u. Show that the image $T_u(S)$ is convex.*

SOLUTION. If $P', Q' \in T_u(S)$, then there exist P and Q in S such that $P + u = P'$ and $Q + u = Q'$. Then

$$tP' + (1-t)Q' = tP + (1-t)Q + u.$$

9. *Let S be a convex set in the vector space V and let c be a number. Let cS denote the set of all elements cv with v in S. Show that cS is convex.*

SOLUTION. The map $L\colon V \to V$ defined by $L(v) = cv$ is linear and $L(S) = cS$, so Exercise 1 concludes the proof.

10. *Let v, w be linearly independent elements of a vector space V. Let $F\colon V \to W$ be a linear map. Assume that $F(v)$, $F(w)$ are linearly dependent. Show that the image under F of the parallelogram spanned by v and w is either a point or a line segment.*

SOLUTION. We have $L(t_1 v + t_2 w) = t_1 L(v) + t_2 L(w)$. There exist numbers a and b that are not both zero such that $aL(v) + bL(w) = O$. Assume that $a \ne 0$. We have $L(v) = cL(w)$ and therefore

$$L(t_1 v + t_2 w) = (ct_1 + t_2)L(w).$$

Conclude.

CHAPTER IV

Linear Maps and Matrices

IV, §1 Linear Map Associated with a Matrix

1. *In each case, find the vector $L_A(X)$.*

(a) $A = \begin{pmatrix} 2 & 1 \\ 1 & 0 \end{pmatrix}$, $X = \begin{pmatrix} 3 \\ -1 \end{pmatrix}$ (b) $A = \begin{pmatrix} 1 & 0 \\ 0 & 0 \end{pmatrix}$, $X = \begin{pmatrix} 5 \\ 1 \end{pmatrix}$

(c) $A = \begin{pmatrix} 1 & 1 \\ 0 & 1 \end{pmatrix}$, $X = \begin{pmatrix} 4 \\ 1 \end{pmatrix}$ (d) $A = \begin{pmatrix} 0 & 0 \\ 0 & 1 \end{pmatrix}$, $X = \begin{pmatrix} 7 \\ -3 \end{pmatrix}$

SOLUTION. (a) $\begin{pmatrix} 5 \\ 3 \end{pmatrix}$ (b) $\begin{pmatrix} 5 \\ 0 \end{pmatrix}$ (c) $\begin{pmatrix} 5 \\ 1 \end{pmatrix}$ (d) $\begin{pmatrix} 0 \\ -3 \end{pmatrix}$.

IV, §2 The Matrix Associated with a Linear Map

1. *Find the matrix associated with the following linear maps.*

(a) $F: \mathbf{R}^4 \to \mathbf{R}^2$ *given by* $F({}^t(x_1, x_2, x_3, x_4)) = {}^t(x_1, x_2)$ *(the projection)*

(b) The projection from $\mathbf{R}^4$ *to* $\mathbf{R}^3$

(c) $F: \mathbf{R}^2 \to \mathbf{R}^2$ *given by* $F({}^t(x, y)) = {}^t(3x, 3y)$

(d) $F: \mathbf{R}^n \to \mathbf{R}^n$ *given by* $F(X) = 7X$

(e) $F: \mathbf{R}^n \to \mathbf{R}^n$ *given by* $F(X) = -X$

(f) $F: \mathbf{R}^4 \to \mathbf{R}^4$ *given by* $F({}^t(x_1, x_2, x_3, x_4)) = {}^t(x_1, x_2, 0, 0)$

SOLUTION. (a) $\begin{pmatrix} 1 & 0 & 0 & 0 \\ 0 & 1 & 0 & 0 \end{pmatrix}$ (b) $\begin{pmatrix} 1 & 0 & 0 & 0 \\ 0 & 1 & 0 & 0 \\ 0 & 0 & 1 & 0 \end{pmatrix}$ (c) $\begin{pmatrix} 3 & 0 \\ 0 & 3 \end{pmatrix}$

(d) $7I = \begin{pmatrix} 7 & 0 & \dots & 0 \\ & & & \\ 0 & \dots & 0 & 7 \end{pmatrix}$ (e) $-I = \begin{pmatrix} -1 & 0 & \dots & 0 \\ & & & \\ 0 & \dots & 0 & -1 \end{pmatrix}$ (f) $\begin{pmatrix} 1 & 0 & 0 & 0 \\ 0 & 1 & 0 & 0 \\ 0 & 0 & 0 & 0 \\ 0 & 0 & 0 & 0 \end{pmatrix}$.

2. *Find the matrix $R(\theta)$ associated with the rotation for each of the following values of θ.*

(a) $\pi/2$ *(b)* $\pi/4$ *(c)* π *(d)* $-\pi$ *(e)* $-\pi/3$ *(f)* $\pi/6$ *(g)* $5\pi/4$

SOLUTION.

(a) $\begin{pmatrix} 0 & -1 \\ 1 & 0 \end{pmatrix}$ (b) $\begin{pmatrix} \sqrt{2}/2 & -\sqrt{2}/2 \\ \sqrt{2}/2 & \sqrt{2}/2 \end{pmatrix}$ (c) $\begin{pmatrix} -1 & 0 \\ 0 & -1 \end{pmatrix}$ (d) $\begin{pmatrix} -1 & 0 \\ 0 & -1 \end{pmatrix}$

(e) $\begin{pmatrix} \frac{1}{2} & \sqrt{3}/2 \\ -\sqrt{3}/2 & \frac{1}{2} \end{pmatrix}$ (f) $\begin{pmatrix} \sqrt{3}/2 & -\frac{1}{2} \\ \frac{1}{2} & \sqrt{3}/2 \end{pmatrix}$ (g) $\begin{pmatrix} -\sqrt{2}/2 & \sqrt{2}/2 \\ -\sqrt{2}/2 & -\sqrt{2}/2 \end{pmatrix}$.

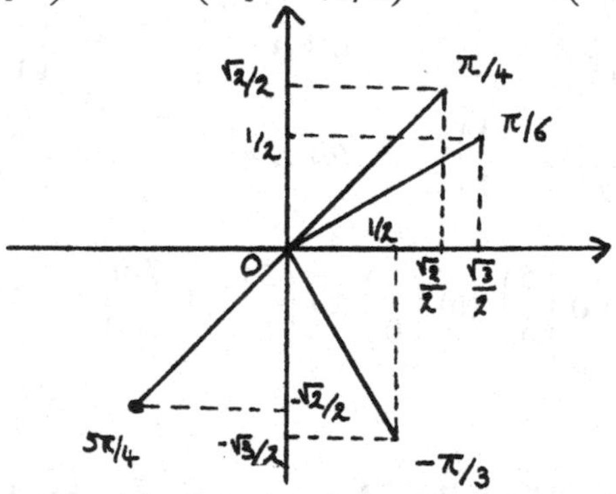

3. *In general, let $\theta > 0$. What is the matrix associated with the rotation by an angle $-\theta$ (i.e. clockwise rotation by θ)?*

SOLUTION. $\begin{pmatrix} \cos\theta & \sin\theta \\ -\sin\theta & \cos\theta \end{pmatrix}$.

4. *Let $X = {}^t(1,2)$ be a point of the plane. Let F be the rotation through an angle of $\pi/4$. What are the coordinates of $F(X)$ relative to the usual basis $\{E_1, E_2\}$?*

SOLUTION. We simply multiply $\begin{pmatrix} \sqrt{2}/2 & -\sqrt{2}/2 \\ \sqrt{2}/2 & \sqrt{2}/2 \end{pmatrix}\begin{pmatrix} 1 \\ 2 \end{pmatrix} = \begin{pmatrix} -\sqrt{2}/2 \\ 3\sqrt{2}/2 \end{pmatrix}$; so the coordinates of $F(X)$ with respect to the usual basis are $(-\sqrt{2}/2, 3\sqrt{2}/2)$.

5. *Same question when* $X={}^t(-1,3)$, *and F is the rotation through* $\pi/2$.

SOLUTION. We have $\begin{pmatrix} 0 & -1 \\ 1 & 0 \end{pmatrix}\begin{pmatrix} -1 \\ 3 \end{pmatrix} = \begin{pmatrix} -3 \\ -1 \end{pmatrix}$; so the coordinates of $F(X)$ with respect to the usual basis are $(-3,-1)$.

6. *Let* $F\colon \mathbf{R}^n \to \mathbf{R}^n$ *be a linear map which is invertible. Show that if A is the matrix associated with F, then* A^{-1} *is the matrix associated with the inverse of F.*

SOLUTION. The assertion follows from the fact that when we compose linear maps we multiply the associated matrices.

7. *Let F be a rotation through an angle* θ. *Show that for any vector X in* $\mathbf{R}^2$ *we have* $\|X\| = \|F(X)\|$ *(i.e. F preserves norms), where* $\|(a,b)\| = \sqrt{a^2+b^2}$.

SOLUTION. If $X=(x,y)$, then

$$\begin{pmatrix} \cos\theta & -\sin\theta \\ \sin\theta & \cos\theta \end{pmatrix}\begin{pmatrix} x \\ y \end{pmatrix} = \begin{pmatrix} x\cos\theta - y\sin\theta \\ x\sin\theta + y\cos\theta \end{pmatrix};$$

so

$$\|F(X)\|^2 = (x\cos\theta - y\sin\theta)^2 + (x\sin\theta + y\cos\theta)^2 = x^2+y^2 = \|X\|^2.$$

8. *Let c be a number, and let* $L\colon \mathbf{R}^n \to \mathbf{R}^n$ *be the linear map such that* $L(X)=cX$. *What is the matrix associated with this linear map?*

SOLUTION. With respect to the usual basis, the desired matrix has the form

$$\begin{pmatrix} c & 0 & \dots & 0 \\ 0 & & & \vdots \\ \vdots & & & 0 \\ 0 & \dots & 0 & c \end{pmatrix} = cI.$$

9. *Let F_θ be rotation by an angle θ. If θ, φ are numbers, compute the matrix of the linear map $F_\theta F_\varphi$ and show that it is the matrix $F_{\theta+\varphi}$.*

SOLUTION. The matrix of $F_\theta F_\varphi$ is given by

$$\begin{pmatrix} \cos\theta & -\sin\theta \\ \sin\theta & \cos\theta \end{pmatrix}\begin{pmatrix} \cos\varphi & -\sin\varphi \\ \sin\varphi & \cos\varphi \end{pmatrix} = \begin{pmatrix} \cos(\theta+\varphi) & -\sin(\theta+\varphi) \\ \sin(\theta+\varphi) & \cos(\theta+\varphi) \end{pmatrix}.$$

We use the trigonometric formulas

$$\cos\theta\cos\varphi - \sin\theta\sin\varphi = \cos(\theta+\varphi)$$

and

$$\cos\theta\sin\varphi + \sin\theta\cos\varphi = \sin(\theta+\varphi).$$

10. *Let F_θ be rotation by an angle θ. Show that F_θ is invertible, and determine the matrix associated with F_θ^{-1}.*

SOLUTION. It is clear from Exercise 9 that F_θ is invertible and that $F_\theta^{-1} = F_{-\theta}$ because $F_0 = \text{id}$. So the matrix associated with F_θ^{-1} is

$$\begin{pmatrix} \cos\theta & \sin\theta \\ -\sin\theta & \cos\theta \end{pmatrix}.$$

IV, §3 Bases, Matrices, and Linear Maps

1. *In each of the following cases, find $M_B^{B'}(\text{id})$. The vector space in each case is $\mathbf{R}^3$.*

(a) $B = \{(1,1,0), (-1,1,1), (0,1,2)\}$
$B' = \{(2,1,1), (0,0,1), (-1,1,1)\}$

(b) $B = \{(3,2,1),(0,-2,5),(1,1,2)\}$
$B' = \{(1,1,0),(-1,2,4),(2,-1,1)\}$

SOLUTION. (a) $\begin{pmatrix} \frac{2}{3} & 0 & \frac{1}{3} \\ -1 & 0 & 1 \\ \frac{1}{3} & 1 & \frac{2}{3} \end{pmatrix}$ (b) $\begin{pmatrix} \frac{11}{5} & \frac{-11}{5} & \frac{3}{5} \\ \frac{2}{15} & \frac{13}{15} & \frac{2}{5} \\ \frac{7}{15} & \frac{23}{15} & \frac{2}{5} \end{pmatrix}$.

2. *Let $L\colon V \to V$ be a linear map. Let $B = \{v_1, \ldots, v_n\}$ be a basis of V. Suppose that there are numbers $c_1, \ldots, c_n$ such that $L(v_i) = c_i v_i$ for $i = 1, \ldots, n$. What is $M_B^B(L)$?*

SOLUTION. We have $M_{\mathscr{B}'}^{\mathscr{B}}(L) = \begin{pmatrix} c_1 & 0 & \ldots & 0 \\ 0 & c_2 & & : \\ : & & & 0 \\ 0 & \ldots & 0 & c_n \end{pmatrix}$.

3. *For each real number θ, let $F_\theta\colon \mathbf{R}^2 \to \mathbf{R}^2$ be the linear map represented by the matrix*

$$R(\theta) = \begin{pmatrix} \cos\theta & -\sin\theta \\ \sin\theta & \cos\theta \end{pmatrix}.$$

Show that if θ, θ' are real numbers then, $F_\theta F_{\theta'} = F_{\theta+\theta'}$. (You must use the addition formula for sine and cosine.) Also show that $F_\theta^{-1} = F_{-\theta}$.

SOLUTION. See Exercises 9 and 10 in §2.

4. *In general, let $\theta > 0$. What is the matrix associated with the rotation by an angle $-\theta$ (i.e. clockwise rotation by θ)?*

SOLUTION. $\begin{pmatrix} \cos\theta & \sin\theta \\ -\sin\theta & \cos\theta \end{pmatrix}$.

5. *Let $X = {}^t(1,2)$ be a point of the plane. Let F be the rotation through an angle of $\pi/4$. What are the coordinates of $F(X)$ relative to the usual basis $\{E_1, E_2\}$?*

SOLUTION. See Exercise 4 in §2.

6. *Same question when* $X={}^t(-1,3)$, *and F is the rotation through* $\pi/2$.

SOLUTION. See Exercise 5 in §2.

7. *In general, let F be the rotation through an angle* θ. *Let* (x,y) *be a point of the plane in the standard coordinate system. Let* (x',y') *be the coordinates of the point in the rotates system. Express* x', y' *in terms of x, y and* θ.

SOLUTION. Let $E_1'=F(E_1)$ and $E_2'=F(E_2)$. We have

$$xE_1+yE_2=x'E_1'+y'E_2',$$

so

$$F^{-1}(xE_1+yE_2)=F^{-1}(x'E_1'+y'E_2')=x'E_1+y'E_2.$$

Therefore Exercise 4 implies

$$\begin{pmatrix}\cos\theta & \sin\theta\\ -\sin\theta & \cos\theta\end{pmatrix}\begin{pmatrix}x\\ y\end{pmatrix}=\begin{pmatrix}x'\\ y'\end{pmatrix},$$

hence $x'=x\cos\theta+y\sin\theta$ and $y'=-x\sin\theta+y\cos\theta$.

8. *In each of the following cases, let* $D=d/dt$ *be the derivative. We give a set of linearly independent functions B. These generate a vector space V, and D is a linear map from V into itself. Find the matrix associated with D relative to the basis B, B.*

(a) $\{e^t, e^{2t}\}$
(b) $\{1, t\}$
(c) $\{e^t, te^t\}$
(d) $\{1, t, t^2\}$
(e) $\{1, t, e^t, e^{2t}, te^{2t}\}$
(f) $\{\sin t, \cos t\}$

SOLUTION.

(a) $\begin{pmatrix}1&0\\0&2\end{pmatrix}$ (b) $\begin{pmatrix}0&1\\0&0\end{pmatrix}$ (c) $\begin{pmatrix}1&1\\0&1\end{pmatrix}$ (d) $\begin{pmatrix}0&1&0\\0&0&2\\0&0&0\end{pmatrix}$ (e) $\begin{pmatrix}0&1&0&0&0\\0&0&0&0&0\\0&0&1&0&0\\0&0&0&2&1\\0&0&0&0&2\end{pmatrix}$

(f) $\begin{pmatrix} 0 & -1 \\ 1 & 0 \end{pmatrix}$.

9. *(a) Let N be a square matrix. We say that N is nilpotent if there exists a positive integer r such that $N^r = O$. Prove that if N is nilpotent then $I - N$ is invertible.*
(b) State and prove the analogous statement for linear maps of a vector space into itself.

SOLUTION. (a) See Exercise 37 in §3 of Chapter II.

(b) If there exists a positive integer r such that $L^r = O$, then $I - L$ is invertible and its inverse is given by $I + L + L^2 + \ldots + L^{r-1}$. The proof consists of verifying that

$$(I - L)(I + L + L^2 + \ldots + L^{r-1}) = (I - L)(I + L + L^2 + \ldots + L^{r-1}) = I.$$

10. *Let P_n denote the vector space of polynomials of degree $\leq n$. Then the derivative D: $P_n \to P_n$ is a linear map of P_n into itself. Let I be the identity mapping. Prove that the following linear maps are invertible:*
(a) $I - D^2$.
(b) $D^m - I$ for any positive integer m.
(c) $D^m - cI$ for any number $c \neq 0$.

SOLUTION. For all integers $p \geq n+1$ we have $D^p = O$, and if a is a number and q a positive integer, then aD^q is nilpotent because

$$(aD^q)^{n+1} = a^{n+1}(D^{n+1})^q = 0,$$

so by Exercise 9 we see that:

(a) The map $I - D^2$ is invertible.

(b) The map $I - D^m$ is invertible for all positive integers m.

(b) The map $\frac{1}{c}D^m$ is nilpotent, so $\frac{1}{c}D^m - I$ is invertible and therefore the map $D^m - cI$ is invertible for any number $c \neq 0$.

11. *Let A be the $n \times n$ matrix which is upper triangular, with zeros on the diagonal, 1 just above the diagonal, and zeros elsewhere.*
(a) How would you describe the effect of L_A on the standard basis vectors $\{E^1,,..E^n\}$ of K^n?

(b) Show that $A^n = O$ and $A^{n-1} = O$ y using the effect of powers of A on the basis vectors.

SOLUTION. (a) We see that $AE^1 = O$ and that if $j \geq 2$, then $AE^j = E^{j-1}$ because as we see from the disposition

$$\begin{pmatrix} 0 & 1 & 0 & \dots & 0 \\ 0 & 0 & 1 & & \vdots \\ \vdots & \vdots & & & 0 \\ 0 & 0 & \dots & 0 & 1 \\ 0 & 0 & 0 & \dots & 0 \end{pmatrix} \begin{pmatrix} 0 \\ \vdots \\ 1 \\ \vdots \\ 0 \end{pmatrix},$$

only the product of the $j-1$ row of A with E^j is not 0 and is in fact 1.

(b) Induction and (a) show that if $1 \leq p \leq n$, then

$$A^p E^1 = A^p E^2 = \dots = A^p E^p = O \quad \text{and} \quad A^p E^{p+j} = E^j$$

for $1 \leq j \leq n-p$. So letting $p = n-1$, we see that the matrix A^{n-1} is

$$A^{n-1} = \begin{pmatrix} 0 & 0 & \dots & 0 & 1 \\ 0 & 0 & \dots & & 0 \\ \vdots & \vdots & & & \vdots \\ 0 & 0 & \dots & & 0 \\ 0 & 0 & 0 & \dots & 0 \end{pmatrix}$$

so that $A^n = O$.

CHAPTER V

Scalar Products and Orthogonality

V, §1 Scalar Products

1. *Let V be a vector space with a scalar product. Show that $\langle O, v\rangle = 0$ for all v in V.*

SOLUTION. We have $\langle O, v\rangle = \langle v - v, v\rangle = \langle v, v\rangle - \langle v, v\rangle = 0$.

2. *Assume that the scalar product is positive definite. Let $v_1, \ldots, v_n$ be non-zero elements which are mutually perpendicular, that is $\langle v_i, v_j\rangle = 0$ if $i \neq j$. Show that they are linearly independent.*

SOLUTION. Suppose that $a_1v_1 + \ldots + a_nv_n = O$ for some scalars $a_1, \ldots, a_n$. Then

$$0 = \langle v_j, a_1v_1 + \ldots + a_nv_n\rangle = a_1\langle v_j, v_1\rangle + \ldots + a_j\langle v_j, v_j\rangle + a_n\langle v_j, v_n\rangle = a_j\langle v_j, v_j\rangle,$$

and, since the scalar product is positive definite, we conclude that $a_j = 0$.

3. *Let M be a square $n \times n$ matrix which is equal to its transpose. If X, Y are column n-vectors, then ${}^tXMY$ is a 1×1 matrix which we identify with a number. Show that the map $(X, Y) \mapsto XMY$ satisfies the three properties **SP 1, SP 2, SP 3**. Give an example of a 2×2 matrix M such that the product is not positive definite.*

SOLUTION. This exercise is simply Exercise 10 in §3 of Chapter II.

V, §2 Orthogonal Bases, Positive Definite Case

0. *What is the dimension of the subspace of* $\mathbf{R}^6$ *perpendicular to the two vectors* $(1,1,-2,3,4,5)$ *and* $(0,0,1,1,0,7)$*?*

SOLUTION. The two given vectors are linearly independent, so the answer is $4=6-2$.

Remark: In general, the answers to Exercises 1, 2, 4, 5, and 6 are not unique.

1. *Find an orthonormal basis for the subspace of* $\mathbf{R}^3$ *generated by the following vectors:*
(a) $(1,1,-1)$ *and* $(1,0,1)$ *(b)* $(2,1,1)$ *and* $(1,3,-1)$

SOLUTION. (a) Let $A=(1,1,-1)$ and $B=(1,0,1)$; then $B\cdot A=0$ so since $\|A\|=\sqrt{3}$ and $\|B\|=\sqrt{2}$, we see that one possible answer is

$$\left\{\frac{1}{\sqrt{3}}(1,1,-1),\frac{1}{\sqrt{2}}(1,0,1)\right\}.$$

(b) Let $A=(2,1,1)$ and $B=(1,3,-1)$. Then $B\cdot A=4$ and $A\cdot A=6$ so that

$$B'=B-\frac{B\cdot A}{A\cdot A}A=\tfrac{1}{3}(-1,7,-5).$$

Normalizing our orthogonal set of vectors, we see that one possible answer is

$$\left\{\frac{1}{\sqrt{6}}(2,1,1),\frac{1}{\sqrt{75}}(-1,7,-5)\right\}.$$

2. *Find an orthonormal basis for the subspace of* $\mathbf{R}^4$ *generated by the following vectors:*
(a) $(1,2,1,0)$ *and* $(1,2,3,1)$
(b) $(1,1,0,0)$, $(1,-1,1,1)$, *and* $(-1,0,2,1)$

SOLUTION. (a) Let $A=(1,2,1,0)$ and $B=(1,2,3,1)$. Then $B\cdot A=8$ and $A\cdot A=6$, so that

$$B' = B - \frac{B \cdot A}{A \cdot A} A = \tfrac{1}{3}(-1, -2, 5, 3).$$

Normalizing our orthogonal set of vectors, we see that one possible answer is

$$\left\{ \frac{1}{\sqrt{6}}(1, 2, 1, 0), \frac{1}{\sqrt{39}}(-1, -2, 5, 3) \right\}.$$

(b) Let $A = (1, 1, 0, 0)$, $B = (1, -1, 1, 1)$, and $C = (-1, 0, 2, 1)$. Note that $A \cdot B = 0$, so $B' = B$. For C' we find

$$C' = C - \frac{C \cdot A}{A \cdot A} A - \frac{C \cdot B'}{B' \cdot B'} B' = \tfrac{1}{2}(-2, 2, 3, 1).$$

Normalizing our orthogonal set of vectors, we see that one possible answer is

$$\left\{ \frac{1}{\sqrt{2}}(1, 1, 0, 0), \frac{1}{2}(1, -1, 1, 1), \frac{1}{\sqrt{18}}(-2, 2, 3, 1) \right\}.$$

3. *In Exercises 3 through 5 we consider the vector space of continuous real values functions on the interval* $[0, 1]$. *We define the scalar product of two such functions f, g by the rule*

$$\langle f, g \rangle = \int_0^1 f(t) g(t)\, dt.$$

Using standard properties of the integral, verify that this is a scalar product.

SOLUTION. Since $f(t)g(t) = g(t)f(t)$, **SP 1** holds. For **SP 2** and **SP 3**, we have

$$\int_0^1 f(g + h)\, dt = \int_0^1 fg + fh\, dt = \int_0^1 fg\, dt + \int_0^1 fh\, dt$$

and

$$\int_0^1 (cf) g\, dt = c \int_0^1 fg\, dt = \int_0^1 f(cg)\, dt.$$

4. *Let V be the subspace of functions generated by the two functions f, g such that* $f(t) = t$ *and* $g(t) = t^3$. *Find an orthonormal basis for V.*

SOLUTION. Since $\langle f, g\rangle = 1/4$ and $\langle f, f\rangle = 1/3$, we see that

$$\tilde{g} = g - \frac{\langle f, g\rangle}{\langle f, f\rangle} f = t^2 - \tfrac{3}{4}t.$$

Since $\|\tilde{g}\| = 1/\sqrt{80}$, we see that one solution is

$$\left\{\sqrt{3}t, \sqrt{80}\left(t^2 - \tfrac{3}{4}t\right)\right\}.$$

5. *Let V be the subspace of functions generated by the three functions* $1, t, t^2$ *(where 1 is the constant function). Find an orthonormal basis for V.*

SOLUTION. Name the three functions f, g, and h, respectively. Then we have

$$\tilde{g} = g - \frac{\langle f, g\rangle}{\langle f, f\rangle} f = t - \tfrac{1}{2}$$

and

$$\tilde{h} = h - \frac{\langle h, f\rangle}{\langle f, f\rangle} f - \frac{\langle h, \tilde{g}\rangle}{\langle \tilde{g}, \tilde{g}\rangle} \tilde{g} = t^2 - t + \tfrac{1}{6}.$$

Normalizing our orthogonal set of vectors, we see that one possible answer is

$$\left\{1, \sqrt{12}\left(t - \tfrac{1}{2}\right), \sqrt{180}\left(t^2 - t + \tfrac{1}{6}\right)\right\}.$$

6. *Find an orthonormal basis for the subspace of* $\mathbf{C}^3$ *generated by the following vectors:*
(a) $(1, i, 0)$ *and* $(1, 1, 1)$ *(b)* $(1, -1, -i)$ *and* $(i, 1, 2)$

SOLUTION. (a) Let $A = (1, i, 0)$ and $B = (1, 1, 1)$. Then $\langle B, A\rangle = 1 - i$ and $\langle A, A\rangle = 2$, so

$$B' = B - \frac{\langle B, A\rangle}{\langle A, A\rangle} A = \tfrac{1}{2}(1 + i, 1 - i, 2).$$

Therefore a solution is

$$\left\{\frac{1}{\sqrt{2}}(1,i,0), \frac{1}{\sqrt{8}}(1+i,1-i,2)\right\}.$$

(b) Let $A=(1,-1,i)$ and $B=(i,1,2)$. Then $\langle B,A\rangle=-1+3i$ and $\langle A,A\rangle=3$, so

$$B'=B-\frac{\langle B,A\rangle}{\langle A,A\rangle}A=\tfrac{1}{3}(1,2+3i,3-i).$$

Hence a solution is given by

$$\left\{\frac{1}{\sqrt{3}}(1,-1,i), \frac{1}{\sqrt{24}}(1,2+3i,3-i)\right\}.$$

7. *(a) Let V be the vector space of all* $n \times n$ *matrices over* $\mathbf{R}$, *and define the scalar product of two matrices A, B by* $\langle A,B\rangle=\operatorname{tr}(AB)$ *where* tr *is the trace (sum of the diagonal elements). Show that this is a scalar product.*

(b) If A is a real symmetric matrix, show that $\operatorname{tr}(AA)\geq 0$, *and* $\operatorname{tr}(AA)>0$ *if* $A\neq O$. *Thus the trace defines a positive definite scalar product on the space of real symmetric matrices.*

(c) Let V be the vector space of real $n\times n$ *symmetric matrices. What is* dim V? *What is the dimension of the subspace W consisting of those matrices A such that* $\operatorname{tr}(A)=0$? *What is the dimension of the orthogonal complement* $W^{\perp}$ *relative to the positive definite scalar product of part (b)?*

SOLUTION. (a) In Exercise 27 in §3 of Chapter 2, we had $\operatorname{tr}(AB)=\operatorname{tr}(BA)$ so the property **SP 1** is verified. Furthermore, $\operatorname{tr}(A+B)=\operatorname{tr}(A)+\operatorname{tr}(B)$ and $\operatorname{tr}(cA)=c\operatorname{tr}(A)$; hence **SP 2** and **SP 3** follow at once.

We contend that this scalar product is non-degenerate. Let E_{kp} be the matrix with all entries 0 except the kp-entry, which is equal to 1. Then if $A=(a_{ij})$ and $\langle A,B\rangle=0$ for all B, we must have $\langle A,E_{kp}\rangle=0$ for all $0\leq k,p\leq n$. But $\operatorname{tr}(AE_{kp})=a_{pk}$. Indeed, suppose $AE_{kp}=(c_{ij})$, then if $m\neq p$, we have

$$c_{mm}=\sum_{r=1}^{n}a_{mr}b_{rm}=0,$$

and

$$c_{pp} = \sum_{r=1}^{n} a_{pr} b_{rp} = a_{pk}.$$

Conclude.

(b) See Exercise 13 in §3 of Chapter III.

(c) (i) In Exercise 6, §1 of Chapter II, we saw that the dimension of the space of symmetric $n \times n$ matrices is $\frac{n(n+1)}{2}$.

(ii) We contend that $\dim W = \frac{n(n+1)}{2} - 1$. Consider the linear map $L\colon V \to \mathbf{R}$ defined by $L(A) = \operatorname{tr}(A)$. Clearly, the image of L is all of $\mathbf{R}$, so the result drops out.

(iii) Since $\dim W + \dim W^{\perp} = \dim V$, we have $\dim W^{\perp} = 1$.

8. *Notation as in Exercise 7, describe the orthogonal complement of the subspace of diagonal matrices. What is the dimension of this orthogonal complement?*

SOLUTION. The dimension of the subspace D of diagonal matrices is n; see Exercise 7 in §1 of Chapter II.

Description of $D^{\perp}$. If $B = (b_{ij}) \in D^{\perp}$, then, given any $A = (a_{ij}) \in D$, we have

$$\langle A, B \rangle = \operatorname{tr}(AB) = \sum_{k=1}^{n} a_{kk} b_{kk} = 0.$$

Taking the scalar product with the diagonal elementary matrices we see that $B = (b_{ij}) \in D^{\perp}$ if and only if $b_{kk} = 0$ for all $1 \le k \le n$. Since $\dim D + \dim D^{\perp} = \dim V$, we conclude that $\dim D^{\perp} = \frac{n(n-1)}{2}$. Of course we see that a basis for D is given by $\{E_{ij} + E_{ji}\}_{1 \le i < j \le n}$.

9. *Let V be a finite dimensional space over $\mathbf{R}$, with a positive definite scalar product. Let $\{v_1, \dots, v_m\}$ be a set of elements of V, or norm 1 and mutually perpendicular (i.e. $\langle v_i, v_j \rangle = 0$ if $i \ne j$). Assume that for every $v \in V$ we have*

$$\| v \|^2 = \sum_{i=1}^{m} \langle v, v_i \rangle^2.$$

Show that $\{v_1,\ldots,v_m\}$ *is basis for V.*

SOLUTION. The set of vectors is orthogonal, so $\{v_1,\ldots,v_m\}$ are linearly independent. It suffices to show that this set generates V. Given v, let $w=\sum_{i=1}^{m}\langle v,v_i\rangle v_i$. Then

$$\langle v-w,v-w\rangle=\|v\|^2-2\langle v,w\rangle+\langle w,w\rangle,$$

but $\langle w,w\rangle=\sum_{i=1}^{m}\langle v,v_i\rangle^2=\langle v,w\rangle=\|v\|^2$, so $\langle v-w,v-w\rangle=0$, proving that $v=\sum_{i=1}^{m}\langle v,v_i\rangle v_i$.

10. *Let V be a finite dimensional vector space over* **R**, *with a positive definite scalar product. Prove the parallelogram law, for any elements* $u,\ w\in V$

$$\|u+v\|^2+\|u-v\|^2=2\left(\|u\|^2+\|v\|^2\right).$$

SOLUTION. The left side of the expression is equal to

$$\langle u+v,u+v\rangle+\langle u-v,u-v\rangle=2\langle u,u\rangle+2\langle v,v\rangle=2\left(\|u\|^2+\|v\|^2\right).$$

V, §3 Applications to Linear Equations; The Rank

1. *Find the rank of the following matrices*

(a) $\begin{pmatrix}2&1&3\\7&2&0\end{pmatrix}$ (b) $\begin{pmatrix}-1&2&-2\\3&4&-5\end{pmatrix}$ (c) $\begin{pmatrix}1&2&7\\2&4&-1\end{pmatrix}$

(d) $\begin{pmatrix}1&2&-3\\-1&-2&3\\4&8&-12\\0&0&0\end{pmatrix}$ (e) $\begin{pmatrix}2&0\\0&-5\end{pmatrix}$ (f) $\begin{pmatrix}-1&0&1\\0&2&3\\0&0&7\end{pmatrix}$

(g) $\begin{pmatrix}2&0&0\\-5&1&2\\3&8&-7\end{pmatrix}$ (h) $\begin{pmatrix}1&2&-3\\-1&-2&3\\4&8&-12\\1&-1&5\end{pmatrix}$.

SOLUTION. (a) 2 (b) 2 (c) 2 (d) 1 (e) 2 (f) 3 (g) 3 (h) 2.

2. *Let A, B be two matrices which can be multiplied. Show that*

$$\text{rank of } AB \leq \text{rank of } A, \textit{ and also } \text{rank of } AB \leq \text{rank of } B.$$

SOLUTION. Let L_A and L_B be the linear maps associated to A and B, respectively.
(i) Then

$$K^n \xrightarrow[L_B]{} K^m \xrightarrow[L_A]{} K^p.$$

If $y \in \text{Im}\,(L_A L_B)$, then there exists $x \in K^n$ such that $L_A(L_B(x)) = y$. Hence $y \in \text{Im}\,(L_A)$, and thus $\text{Im}\,(L_A L_B) \subset \text{Im}\,(L_A)$, so $\text{rank}\, AB \leq \text{rank}\, A$.
(ii) Now consider

$$K^n \xrightarrow[L_B]{} \text{Im}\,(L_B) \xrightarrow[\tilde{L}_A]{} K^p.$$

Here $\tilde{L}_A$ is the restriction of L_A to $\text{Im}\,(L_B)$. Since

$$\dim \text{Im}\,(\tilde{L}_A) \leq \dim \text{Im}\,(L_B) \text{ and } \text{Im}\,(L_A L_B) = \text{Im}\,(\tilde{L}_A),$$

we conclude that $\text{rank}\, AB \leq \text{rank}\, B$.

3. *Let A be a triangular matrix*

$$\begin{pmatrix} a_{11} & a_{12} & \dots & a_{1n} \\ 0 & a_{22} & & a_{2n} \\ \vdots & \vdots & & \vdots \\ 0 & 0 & \dots & a_{nn} \end{pmatrix}.$$

Assume that none of the diagonal elements is equal to 0. What is the rank of A?

SOLUTION. If we let $x_1 A^1 + x_2 A^2 + \dots + x_n A^n = O$, then we get a system of the form

$$\begin{cases} x_1a_{11} + x_2a_{12} + \ldots + x_{n-1}a_{1n-1} + x_na_{1n} = 0 \\ x_2a_{22} + \ldots + x_{n-1}a_{2n-1} + x_na_{2n} = 0 \\ \vdots \\ x_{n-1}a_{n-1n-1} + x_na_{n-1n} = 0 \\ x_na_{nn} = 0. \end{cases}$$

Since $a_{nn} \neq 0$, we see from the last equation that $x_n = 0$. Since $a_{n-1n-1} \neq 0$, we see from the second to last equation that $x_{n-1} = 0$. By induction we see that $x_1 = x_2 = \ldots = x_n = 0$; thus $\operatorname{rank} A = n$.

In Exercises 4 and 5 we let *S* be the space of solutions of the system of linear equations. Note that the solutions to these exercises are not unique, so we give only one possible answer.

4. *Find the dimension of the space of solutions of the following systems of equations. Also find a basis for this space of solutions.*

(a) $\begin{aligned} 2x + y - z &= 0 \\ y + z &= 0 \end{aligned}$ *(b)* $x - y + z = 0$

(c) $\begin{aligned} 4x + 7y - \pi z &= 0 \\ 2x - y + z &= 0 \end{aligned}$ *(d)* $\begin{aligned} x + y + z &= 0 \\ x - y &= 0 \\ y + z &= 0 \end{aligned}$

SOLUTION. (a) The rank of $\begin{pmatrix} 2 & 1 & -1 \\ 0 & 1 & 1 \end{pmatrix}$ is 2, so $\dim S = 1$. A solution is $(1, 1, -1)$, which is therefore also a basis for *S*.

(b) The rank of $(1 \ \ -1 \ \ 1)$ is 1, so $\dim S = 2$. Two linearly independent solutions are $\{(1,1,0), (0,1,1)\}$, which therefore form a basis for *S*.

(c) Since $(4, 7, -\pi)$ and $(2, -1, 1)$ are linearly independent, $\dim S = 1$. The first equation minus twice the second equals

$$9y - (\pi + 2)z = 0,$$

so if we let $z = 9$, we see that $(\frac{\pi}{2} - \frac{7}{2}, \pi + 2, 9)$ is a solution of the original system; so this vector forms a basis for *S*.

(d) The rank of

$$\begin{pmatrix} 1 & 1 & 1 \\ 1 & -1 & 0 \\ 0 & 1 & 1 \end{pmatrix}$$

is equal to 3, so the space of solutions of the system is reduced to the single element $\{O\}$.

5. *What is the dimension of the space of solutions of the following systems of linear equations?*

(a) $\begin{aligned} 2x - 3y + z &= 0 \\ x + y - z &= 0 \end{aligned}$ (b) $\begin{aligned} 2x + 7y &= 0 \\ x - 2y + z &= 0 \end{aligned}$

(c) $\begin{aligned} 2x - 3y + z &= 0 \\ x + y - z &= 0 \\ 3x + 4y &= 0 \\ 5x + y + z &= 0 \end{aligned}$ (d) $\begin{aligned} x + y + z &= 0 \\ 2x + 2y + 2z &= 0 \end{aligned}$

SOLUTION. (a) The rank of the matrix $\begin{pmatrix} 2 & -3 & 1 \\ 1 & 1 & -1 \end{pmatrix}$ is 2, so $\dim S = 1$.

(b) The rank of $\begin{pmatrix} 2 & 7 & 0 \\ 1 & -2 & 1 \end{pmatrix}$ is 2, so $\dim S = 1$.

(c) Let

$$A = \begin{pmatrix} 2 & -3 & 1 \\ 1 & 1 & -1 \\ 3 & 4 & 0 \\ 5 & 1 & 1 \end{pmatrix}.$$

The first row added to the third equals the fourth row, so we find that the rank of the matrix A is 3. Hence $\dim S = 0$, and therefore $S = \{O\}$.

(d) Note that the second equation is twice the first, so we have to solve

$$x + y + z = 0.$$

Therefore $\dim S = 2$.

6. *Let A be a non-zero vector in n-space. Let P be a point in n-space. What is the dimension of the set of solutions of the equation* $X \cdot A = P \cdot A$ *?*

SOLUTION. Let $S_{.}$ be the set of solutions of $X \cdot A = P \cdot A$ $(*)$ and let $S_{..}$ be the set of solution of $X \cdot A = 0$ $(**)$. By Exercise 7 we know that if we have a solution X_0 to $(*)$, then $S_{.} = X_0 + S_{..}$. Since $X = P$ solves $(*)$ and the dimension of $S_{..}$ is $n-1$, we see that $\dim S_{.} = n-1$.

7. *Let* $AX = B$ *be a system of linear equations, where A is an* $m \times n$ *matrix X is an n-vector, and B is an m-vector. Assume that there is one solution* $X = X_0$. *Show that every solution is of the form* $X_0 + Y$, *where Y is a solution of the homogeneous system* $AY = O$, *and conversely any vector of the form* $X_0 + Y$ *is a solution.*

SOLUTION. Suppose X_1 is a solution of $AX_1 = B$. Then $A(X_1 - X_0) = O$, and we can write $X_1 = X_0 + (X_1 - X_0)$. Conversely, suppose $X_2 = X_0 + Y$, where $AY = O$. Then

$$AX_2 = A(X_0) + A(Y) = B.$$

V, §4 Bilinear Maps and Matrices

1. *Let A be an* $n \times n$ *matrix and assume that A is symmetric, i.e.* $A = {}^tA$. *Let* $g_A \colon K^n \times K^n \to K$ *be its associated bilinear map. Show that*

$$g_A(X, Y) = g_A(Y, X)$$

for all $X, Y \in K^n$, *and thus that* g_A *is a scalar product, i.e. it satisfies conditions* ***SP 1***, ***SP 2***, *and* ***SP 3***.

SOLUTION. Since $A = {}^tA$, we have

$$g_A(X, Y) = {}^tXAY = {}^t({}^tYAX) = {}^tYAX = g_A(Y, X).$$

See Exercise 10 in §3 of Chapter II.

2. *Conversely assume that A is an* $n \times n$ *matrix such that*

$$g_A(X, Y) = g_A(Y, X)$$

for all X, Y. Show that A is symmetric.

SOLUTION. We know that $g(E^k, E^p) = {}^tE^k A E^p = a_{kp}$. The assumption therefore implies that A is symmetric.

3. *Show that the bilinear maps of* $K^n \times K^m$ *into K form a vector space. More generally, let* $\mathrm{Bil}(U \times V, W)$ *be the set of all bilinear forms of* $U \times V$ *into W. Show that* $\mathrm{Bil}(U \times V, W)$ *is a vector space.*

SOLUTION. Let $f, g \in \mathrm{Bil}(U \times V, W)$. Then we have

$$\begin{aligned}(f+g)(u, v+v') &= f(u, v+v') + g(u, v+v') \\ &= f(u,v) + f(u,v') + g(u,v) + g(u,v') \\ &= (f+g)(u,v) + (f+g)(u,v'),\end{aligned}$$

so $(f+g)$ is linear with respect to the second variable. Similar arguments show that $(f+g)$ and cf are bilinear. Thus $\mathrm{Bil}(U \times V, W)$ is a vector space.

4. *Show that the association* $A \mapsto g_A$ *is an isomorphism between the space of* $m \times n$ *matrices, and the space of bilinear maps of* $K^m \times K^n$ *into K.*

SOLUTION. Let

$$\begin{aligned}\psi\colon \mathrm{Mat}_{m\times n}(K) &\to \mathrm{Bil}(K^m \times K^n, K) \\ A &\mapsto g_A\end{aligned}$$

Theorem 4.1. implies that ψ is surjective and injective, so all we have to prove is the linearity of ψ. This result is a simple consequence of the multiplicative properties of matrices, namely,

$$g_{A+B}(X,Y) = {}^tX(A+B)Y = ({}^tXA + {}^tXB)Y = {}^tXAY + {}^tXBY = g_A(X,Y) + g_B(X,Y),$$

so $\psi(A+B) = \psi(A) + \psi(B)$ and

$$g_{cA}(X,Y) = {}^tX(cA)Y = c\,{}^tXAY = cg_A(X,Y);$$

thus $\psi(cA) = c\psi(A)$, thereby proving the assertion.

5. *Write out in full in terms of coordinates the expression for when A is the following matrix, and X, Y are vectors of the corresponding dimension.*

(a) $\begin{pmatrix} 1 & -3 \\ 4 & 1 \end{pmatrix}$ *(b)* $\begin{pmatrix} 4 & 1 \\ -2 & 5 \end{pmatrix}$

(c) $\begin{pmatrix} -5 & 2 \\ \pi & 7 \end{pmatrix}$ *(d)* $\begin{pmatrix} 1 & 2 & -1 \\ -3 & 1 & 4 \\ 2 & 5 & -1 \end{pmatrix}$

(e) $\begin{pmatrix} -4 & 2 & 1 \\ 3 & 1 & 1 \\ 2 & 5 & 7 \end{pmatrix}$ *(f)* $\begin{pmatrix} -\frac{1}{2} & 2 & -5 \\ 1 & \frac{2}{3} & 4 \\ -1 & 0 & 3 \end{pmatrix}$

SOLUTION. In this exercise one can either redo the computation or use the formula given in the text.

(a) ${}^tXAY = 2x_1y_1 - 3x_1y_2 + 4x_2y_1 + x_2y_2$.

(b) ${}^tXAY = 4x_1y_1 + x_1y_2 - 2x_2y_1 + 5x_2y_2$.

(c) ${}^tXAY = -5x_1y_1 + 2x_1y_2 + \pi x_2y_1 + 7x_2y_2$.

(d) ${}^tXAY =$

$$x_1y_1 + 2x_1y_2 - x_1y_3 + -3x_2y_1 + x_2y_2 + 4x_2y_3 + 2x_3y_1 + 5x_3y_2 - x_3y_3.$$

(e) ${}^tXAY =$

$$-4x_1y_1 + 2x_1y_2 + x_1y_3 + 3x_2y_1 + x_2y_2 + x_2y_3 + 2x_3y_1 + 5x_3y_2 + 7x_3y_3.$$

(f) ${}^tXAY = \frac{-1}{2}x_1y_1 + 2x_1y_2 - 5x_1y_3 + x_2y_1 + \frac{2}{3}x_2y_2 + 4x_2y_3 - x_3y_1 + 3x_3y_3$.

6. *Let*

$$C = \begin{pmatrix} 1 & 2 & 3 \\ -1 & 1 & 1 \\ 1 & 0 & 1 \end{pmatrix}$$

and define $g(X, Y) = {}^tXCY$. *Find two vectors* $X,\ Y \in \mathbf{R}^3$ *such that*

$$g(X, Y) \neq g(Y, X).$$

SOLUTION. Let ${}^tX = (1, 0, 0)$ and ${}^tY = (0, 1, 0)$. Then $g(X, Y) = 2$ and $g(Y, X) = -1$.

V, §5 General Orthogonal Bases

1. *Find orthogonal bases of the subspace of* $\mathbf{R}^3$ *generated by the indicated vectors A, B, with respect to the indicated scalar product, written* $X \cdot Y$.

(a) $A = (1,1,1)$, $B = (1,-1,2)$

$X \cdot Y = x_1y_1 + 2x_2y_2 + x_3y_3$

(b) $A = (1,-1,4)$, $B = (-1,1,3)$

$X \cdot Y = x_1y_1 - 3x_2y_2 + x_1y_3 + y_1x_3 - x_3y_2 - x_2y_3$

SOLUTION. (a) We have $B \cdot A = 1 - 2 + 2 = 1$ and $A \cdot A = 4$, so

$$B' = B - \frac{B \cdot A}{A \cdot A} A = \tfrac{1}{4}(3,-5,7).$$

Therefore, a possible solution is $\{(1,1,1), (3,-5,7)\}$.

(b) We already have $B \cdot A = 0$.

2. *Find an orthogonal base for the space* $\mathbf{C}^2$ *over* $\mathbf{C}$, *if the scalar product is given by* $X \cdot Y = x_1y_1 - ix_2y_1 - ix_1y_2 - 2x_2y_2$.

SOLUTION. Let $A = (1,0)$ and $B = (0,1)$. Then $A \cdot A = 1$ and $B \cdot A = -i$, so

$$B' = B - \frac{B \cdot A}{A \cdot A} A = (i,1).$$

Therefore, $\{(1,0), (i,1)\}$ is an orthogonal base for the space $\mathbf{C}^2$ over $\mathbf{C}$.

3. *Same question as in Exercise 2, if the scalar product is given by*

$$X \cdot Y = x_1y_2 + x_2y_1 + 4x_1y_1.$$

SOLUTION. Let A and B be as in Exercise 2. Then $A \cdot A = 4$ and $B \cdot A = 1$, so

$$B' = B - \frac{B \cdot A}{A \cdot A} A = \tfrac{1}{4}(-1,4).$$

Therefore, $\{(1,0), (-1,4)\}$ is an orthogonal base for the space $\mathbf{C}^2$ over $\mathbf{C}$.

V, §6 The Dual Space and Scalar Products

1. *Let A, B be two linearly independent vectors in* $\mathbf{R}^n$. *What is the dimension of the space perpendicular to both A and B?*

SOLUTION. If "perpendicular" refers to a non-degenerate scalar product, then Theorem 6.4 implies that the space perpendicular to both A and B has dimension $n-2$.

2. *Let A, B be two linearly independent vectors in* $\mathbf{C}^n$. *What is the dimension of the subspace of* $\mathbf{C}^n$ *perpendicular to both A and B? (Perpendicularity refers to the ordinary dot product of vectors in* $\mathbf{C}^n$.*)*

SOLUTION. The dimension of the space perpendicular to both A and B is $n-2$.

3. *Let W be the subspace of* $\mathbf{C}^3$ *generated by the vector* $(1, i, 0)$. *Find a basis of* $W^\perp$ *in* $\mathbf{C}^3$ *(with respect to the ordinary dot product of vectors).*

SOLUTION. The vectors $A = (1, -i, 0)$ and $B = (0, 0, 1)$ are linearly independent and perpendicular to $(1, i, 0)$. Since $W^\perp$ has dimension 2, we conclude that $\{A, B\}$ is a basis for $W^\perp$.

4. *Let V be a vector space of finite dimension n over the field K. Let* φ *be a functional on V, and assume that* $\varphi \neq 0$. *What is the dimension of the kernel of* φ*? Proof?*

SOLUTION. We contend that $\dim \operatorname{Ker} \varphi = n-1$. The map $\varphi: V \to K$ is linear, and since $\varphi \neq 0$, there exists a vector v such that $\varphi(v) \neq 0$. Given $x \in K$, we let $v_x = x(\varphi(v))^{-1} v$, so that

$$\varphi(v_x) = x(\varphi(v))^{-1} \varphi(v) = x.$$

Therefore, $\dim \operatorname{Im} \varphi = 1$ which proves our contention.

5. *Let V be a vector space of finite dimension n over the field K. Let* ψ, φ *be two non-zero functionals on V. Assume that there is no element* $c \in K$, $c \neq 0$ *such that* $\psi = c\varphi$. *Show that* $(\operatorname{Ker} \varphi) \cap (\operatorname{Ker} \psi)$ *has dimension* $n-2$.

SOLUTION. Fix a basis $\{v_1, \ldots, v_n\}$ for V, and let A and B be the unique elements of V such that

$$\psi(v) = A \cdot X_v \quad \text{and} \quad \varphi(v) = B \cdot X_v \quad \text{for all } v \text{ in } V,$$

where $\cdot$ is the usual dot product and X_v the coordinates of v. Then the facts that $\psi, \varphi \neq 0$ and there is no constant c such that $\psi = c\varphi$ show that A and B are linearly independent. If W is the subspace generated by A and B, then

$$(\text{Ker}\,\varphi) \cap (\text{Ker}\,\psi) = W^{\perp}.$$

So we see that $\dim\,(\text{Ker}\,\varphi) \cap (\text{Ker}\,\psi) = n - 2$.

6. *Let V be a vector space of finite dimension over the field K. Let V^{**} be the dual space of V^*. Show that each element $v \in V$ gives rise to an element λ_v in V^{**} and that the map $v \mapsto \lambda_v$ gives an isomorphism of V with V^{**}.*

SOLUTION. Fix an element v in V. Then for $\varphi^* \in V^*$ the map $\varphi^* \mapsto \varphi^*(v)$ is linear and is therefore an element of V^{**} which we denote by λ_v. The map $\Phi\colon V \to V^{**}$ defined by $\Phi(v) = \lambda_v$ is linear. Indeed,

$$\lambda_{v_1+v_2}(\varphi^*) = \varphi^*(v_1 + v_2) = \varphi^*(v_1) + \varphi^*(v_2) = \lambda_{v_1}(\varphi^*) + \lambda_{v_2}(\varphi^*),$$

so $\Phi(v_1 + v_2) = \Phi(v_1) + \Phi(v_2)$, and, similarly, we find that $\Phi(cv) = c\Phi(v)$. Since $\dim V = \dim V^* = \dim V^{**}$, all we have to show is that $\text{Ker}\,\Phi = \{O\}$. Suppose $\Phi(v) = O$. Then for all $\varphi^* \in V^*$ we have $\varphi^*(v) = 0$. Selecting a basis for V and considering the coordinate functions, one sees that we must have $v = O$. Therefore, $v \mapsto \lambda_v$ gives an isomorphism of V with V^{**}.

7. *Let V be a finite dimensional vector space over the field K, with a non-degenerate scalar product. Let W be a subspace. Show that $W^{\perp\perp} = W$.*

SOLUTION. Theorem 6.4 implies

$$\dim W + \dim W^{\perp} = \dim V \quad \text{and} \quad \dim W^{\perp} + \dim W^{\perp\perp} = \dim V;$$

therefore, $\dim W^{\perp\perp} = \dim W$. It is now sufficient to prove that W is a subspace of $W^{\perp\perp}$. If $v \in W$, then for all $\overline{w} \in W^{\perp}$ we have $\langle v, \overline{w} \rangle = 0$; so by definition $v \in W^{\perp\perp}$, consequently $W = W^{\perp\perp}$.

V, §7 Quadratic Forms

1. *Let V be a finite dimensional vector space over the field K. Let $f: V \to K$ be a function, and assume that the function g defined by*

$$g(v, w) = f(v + w) - f(v) - f(w)$$

is bilinear. Assume that $f(av) = a^2 f(v)$ for all $v \in V$ and $a \in K$. Show that f is a quadratic form, and determine a bilinear form from which it comes. Show that this bilinear form is unique.

SOLUTION. We have

$$g(v, v) = f(2v) - 2f(v) = 4f(v) - 2f(v) = 2f(v).$$

So let $\tilde{g} = g/2$. Then $\tilde{g}$ is bilinear and the quadratic form it determines is f. Suppose that f is also the quadratic form determined by a bilinear form $\tilde{g}_0$. Then by the formulas given in the text we see that

$$\tilde{g}_0(v, w) = \tfrac{1}{2}[f(v + w) - f(v) - f(w)] = \tilde{g}(v, w),$$

so $\tilde{g}$ is uniquely determined.

2. *What is the associated matrix of the quadratic form $f(X) = x^2 - 3xy + 4y^2$ if ${}^tX = (x, y, z)$?*

SOLUTION. $\begin{pmatrix} 1 & -\frac{3}{2} & 0 \\ \frac{3}{2} & 4 & 0 \\ 0 & 0 & 0 \end{pmatrix}$.

3. *Let x_1, x_2, x_3, x_4 be the coordinates of a vector X, and y_1, y_2, y_3, y_4 the coordinates of a vector Y. Express in terms of these coordinates the bilinear form associated with the following quadratic forms.*
(a) x_1x_2 *(b)* $x_1x_3 + x_4^2$ *(c)* $2x_1x_2 - x_3x_4$ *(d)* $x_1^2 - 5x_2x_3 + x_4^2$

SOLUTION. First we give the matrix associated with the quadratic form, and then we give the expression of the symmetric bilinear map:

(a) $C=\begin{pmatrix} 0 & \frac{1}{2} & 0 & 0 \\ \frac{1}{2} & 0 & 0 & 0 \\ 0 & 0 & 0 & 0 \\ 0 & 0 & 0 & 0 \end{pmatrix}$, so that $g(X,Y)=\frac{1}{2}x_1y_2+\frac{1}{2}x_2y_1$.

(b) $C=\begin{pmatrix} 0 & 0 & \frac{1}{2} & 0 \\ 0 & 0 & 0 & 0 \\ \frac{1}{2} & 0 & 0 & 0 \\ 0 & 0 & 0 & 1 \end{pmatrix}$, so that $g(X,Y)=\frac{1}{2}x_1y_3+\frac{1}{2}x_3y_1+x_4y_4$.

(c) $C=\begin{pmatrix} 0 & 1 & 0 & 0 \\ 1 & 0 & 0 & 0 \\ 0 & 0 & 0 & -\frac{1}{2} \\ 0 & 0 & -\frac{1}{2} & 0 \end{pmatrix}$, so that $g(X,Y)=x_1y_2+x_2y_1-\frac{1}{2}x_3y_4-\frac{1}{2}x_4y_3$.

(d) $C=\begin{pmatrix} 1 & 0 & 0 & 0 \\ 0 & 0 & -\frac{5}{2} & 0 \\ 0 & -\frac{5}{2} & 0 & 0 \\ 0 & 0 & 0 & 1 \end{pmatrix}$, so that $g(X,Y)=x_1y_1-\frac{5}{2}x_2y_3-\frac{5}{2}x_3y_2+x_4y_4$.

4. *Show that if f_1 is the quadratic form of the bilinear form g_1, and f_2 the quadratic form of the bilinear from g_2, then f_1+f_2 is the quadratic form of the bilinear form g_1+g_2.*

SOLUTION. We have

$$(f_1+f_2)(v)=f_1(v)+f_2(v)=g_1(v,v)+g_2(v,v)=(g_1+g_2)(v,v).$$

V, §8 Sylvester's Theorem

1. *Determine the index of nullity and index of positivity for each product determined by the following symmetric matrices, on $\mathbf{R}^2$.*

(a) $\begin{pmatrix} 1 & 2 \\ 2 & -1 \end{pmatrix}$ *(b)* $\begin{pmatrix} 1 & 1 \\ 1 & 1 \end{pmatrix}$ *(c)* $\begin{pmatrix} 1 & -3 \\ -3 & 2 \end{pmatrix}$

SOLUTION. (a) An orthogonal basis is given by $A = (1, 0)$ and $B = (-2, 1)$. Then we have

$$\langle A, A \rangle = 1 \quad \text{and} \quad \langle B, B \rangle = -5,$$

so the index of nullity of the form is 0, and the index of positivity of the scalar product is 1.

(b) An orthogonal basis is given by $A = (1, 0)$ and $B = (-1, 1)$. Then we have

$$\langle A, A \rangle = 1 \quad \text{and} \quad \langle B, B \rangle = 0$$

so the index of nullity of the form is 1, and the index of positivity of the scalar product is 1.

(c) An orthogonal basis is given by $A = (1, 0)$ and $B = (3, 1)$. Then

$$\langle A, A \rangle = 1 \quad \text{and} \quad \langle B, B \rangle = -7,$$

so the index of nullity of the form is 0, and the index of positivity of the scalar product is 1.

2. *Let V be a finite dimensional vector space over* $\mathbf{R}$, *and let $\langle \, , \rangle$ be a scalar product on V. Show that V admits a direct sum decomposition*

$$V = V^+ \oplus V^- \oplus V_0,$$

where V_0 is defined as in Theorem 6.1, and where the product is positive definite on V^+ and negative definite on V^-. Show that the dimensions of the spaces V^+, V^- are the same in all such decompositions.

SOLUTION. Let $\{v_1, \ldots, v_n\}$ be an orthogonal basis for V indexed such that

$$\begin{aligned} \langle v_i, v_i \rangle > 0 & \text{ if } 1 \le i \le r \\ \langle v_i, v_i \rangle < 0 & \text{ if } r+1 \le i \le s \\ \langle v_i, v_i \rangle = 0 & \text{ if } s+1 \le i \le n. \end{aligned}$$

Let V^+ be the space generated by $\{v_1, \ldots, v_r\}$ and V^- the space generated by $\{v_{r+1}, \ldots, v_s\}$. By Theorem 8.1, we know that $\{v_{s+1}, \ldots, v_n\}$ is a basis for V_0. Hence

$$V = V^+ \oplus V^- \oplus V_0.$$

The dimension of V^+ is equal to the index of positivity of the product, and the dimension of V^- is equal to the index of negativity of the product.

We contend that the product is positive definite on V^+. Let $c_i = \langle v_i, v_i \rangle$, and suppose that $v = x_1v_1 + \ldots + x_rv_r$. Then

$$\langle v, v \rangle = \langle x_1v_1 + \ldots + x_rv_r, x_1v_1 + \ldots + x_rv_r \rangle = x_1^2c_1 + \ldots + x_r^2c_r \geq 0$$

and $= 0$ if and only if $v = O$. Similarly, prove that the product is negative definite on V^-.

3. *Let V be a vector space over* $\mathbf{R}$ *of* 2×2 *real symmetric matrices.*

(a) Given a symmetric matrix $A = \begin{pmatrix} x & y \\ y & z \end{pmatrix}$ *show that* (x, y, z) *are the coordinates of A with respect to some basis of the vector space of all* 2×2 *real symmetric matrices. Which basis?*

(b) Let $f(A) = xz - yy = xz - y^2$. *If we view* (x, y, z) *as the coordinates of A then we see that f is a quadratic form on V. Note that* $f(A)$ *is the determinant of A.*

Let W be the subspace of V consisting of all A such that $\operatorname{tr}(A) = 0$. *Show that for* $A \in W$ *and* $A \neq O$ *we have* $f(A) < 0$. *This means that the quadratic form is negative definite on W.*

SOLUTION. (a) Consider the standard basis for the space of 2×2 symmetric matrices, namely,

$$E_1 = \begin{pmatrix} 1 & 0 \\ 0 & 0 \end{pmatrix}, \quad E_2 = \begin{pmatrix} 0 & 1 \\ 1 & 0 \end{pmatrix}, \quad \text{and} \quad E_3 = \begin{pmatrix} 0 & 0 \\ 0 & 1 \end{pmatrix}.$$

Then $A = xE_1 + yE_2 + zE_3$.

(b) Since $\operatorname{tr}(A) = 0$, we have $x = -z$, and thus $xz = -z^2 \leq 0$, so

$$f(A) = xz - y^2 = -(z^2 + y^2) \leq 0.$$

If $f(A) = 0$, then clearly $A = O$, so the quadratic form is negative definite.

CHAPTER VI

Determinants

VI, §2 Existence of Determinants

1. *Let c be a number and let A be a* 3×3 *matrix. Show that*

$$D(cA) = c^3 D(A).$$

SOLUTION. See Exercise 2.

2. *Let c be a number and let A be a* $n\times n$ *matrix. Show that*

$$D(cA) = c^n D(A).$$

SOLUTION. Let $A^1,\dots,A^n$ be the columns of A. Then

$$D(cA) = D(cA^1,\dots,cA^n).$$

The properties of the determinant imply that

$$\begin{aligned} D(cA) &= cD(A^1, cA^2, cA^3,\dots,cA^n) \\ &= c^2 D(A^1, A^2, cA^3, cA^4,\dots,cA^n) \\ &= c^n D(A^1, A^2, A^3, A^4,\dots,A^n), \end{aligned}$$

hence $D(cA) = c^n D(A)$.

VI, §3 Additional Properties of Determinants

1. *Compute the following determinants.*

$$(a)\begin{vmatrix}2&1&2\\0&3&-1\\4&1&1\end{vmatrix}\quad (b)\begin{vmatrix}3&-1&5\\-1&2&1\\-2&4&3\end{vmatrix}\quad (c)\begin{vmatrix}2&4&3\\-1&3&0\\0&2&1\end{vmatrix}\quad (d)\begin{vmatrix}1&2&-1\\0&1&1\\0&2&7\end{vmatrix}$$

$$(e)\begin{vmatrix}-1&5&3\\4&0&0\\2&7&8\end{vmatrix}\quad (f)\begin{vmatrix}3&1&2\\4&5&1\\-1&2&-3\end{vmatrix}.$$

SOLUTION.
(a) –20 (b) 5 (c) 4 (d) 5 (e) –76 (f) –14.

2. *Compute the following determinants.*

$$(a)\begin{vmatrix}1&1&-2&4\\0&1&1&3\\2&-1&1&0\\3&1&2&5\end{vmatrix}\quad (b)\begin{vmatrix}-1&1&2&0\\0&3&2&1\\0&4&1&2\\3&1&5&7\end{vmatrix}\quad (c)\begin{vmatrix}3&1&1\\2&5&5\\8&7&7\end{vmatrix}\quad (d)\begin{vmatrix}4&-9&2\\4&-9&2\\3&1&0\end{vmatrix}$$

$$(e)\begin{vmatrix}4&-1&1\\2&0&0\\1&5&7\end{vmatrix}\quad (f)\begin{vmatrix}2&0&0\\1&1&0\\8&5&7\end{vmatrix}\quad (g)\begin{vmatrix}4&0&0\\0&1&0\\0&0&27\end{vmatrix}\quad (h)\begin{vmatrix}5&0&0\\0&3&0\\0&0&9\end{vmatrix}$$

$$(i)\begin{vmatrix}2&-1&4\\3&1&5\\1&2&3\end{vmatrix}.$$

SOLUTION.
(a) –18 (b) 45 (c) 0 (d) 0 (e) 24 (f) 14 (g) 108

(h) 135 (i) 10.

3. *In general what is the determinant of a diagonal matrix?*

SOLUTION. Expanding according to the first row, we see that

$$\begin{vmatrix}a_{11}&0&\dots&0\\0&a_{22}&&\vdots\\\vdots&&&0\\0&\dots&0&a_{nn}\end{vmatrix}=a_{11}\begin{vmatrix}a_{22}&0&\dots&0\\0&&&\vdots\\\vdots&&&0\\0&\dots&0&a_{nn}\end{vmatrix}=\dots=a_{11}a_{22}\cdots a_{nn}.$$

4. *Compute the determinant* $\begin{vmatrix} \cos\theta & -\sin\theta \\ \sin\theta & \cos\theta \end{vmatrix}$.

SOLUTION. We have

$$\begin{vmatrix} \cos\theta & -\sin\theta \\ \sin\theta & \cos\theta \end{vmatrix} = \cos^2\theta + \sin^2\theta = 1.$$

5. *(a) Let x_1, x_2, x_3 be numbers. Show that*

$$\begin{vmatrix} 1 & x_1 & x_1^2 \\ 1 & x_2 & x_2^2 \\ 1 & x_3 & x_3^2 \end{vmatrix} = (x_2 - x_1)(x_3 - x_1)(x_3 - x_2).$$

(b) If $x_1, \ldots, x_n$ are numbers, then show by induction that

$$\begin{vmatrix} 1 & x_1 & \ldots & x_1^{n-1} \\ 1 & x_2 & \ldots & x_2^{n-1} \\ & & \ldots & \\ 1 & x_n & \ldots & x_n^{n-1} \end{vmatrix} = \prod_{i<j} (x_j - x_i),$$

the symbol on the right meaning that it is the product of all terms $x_j - x_i$ with $i < j$ and i, j integers from 1 to n. This determinant is called the Vandermonde determinant V_n. To do the induction easily, multiply each column by x_1 and subtract it from the next column on the right, starting from the right-hand side. You will find that

$$V_n = (x_n - x_1) \cdots (x_2 - x_1) V_{n-1}.$$

SOLUTION. (a) Expanding according to the first row, we get

$$D = \begin{vmatrix} 1 & x_1 & x_1^2 \\ 1 & x_2 & x_2^2 \\ 1 & x_3 & x_3^2 \end{vmatrix} = \begin{vmatrix} x_2 & x_2^2 \\ x_3 & x_3^2 \end{vmatrix} - x_1 \begin{vmatrix} 1 & x_2^2 \\ 1 & x_3^2 \end{vmatrix} + x_1^2 \begin{vmatrix} 1 & x_2 \\ 1 & x_3 \end{vmatrix}$$

so

$$\begin{aligned} D &= x_2 x_3 (x_3 - x_2) - x_1 (x_3^2 - x_2^2) + x_1^2 (x_3 - x_2) \\ &= (x_3 - x_2)(x_2 - x_1)(x_3 - x_1). \end{aligned}$$

(b) The result is true when $n = 3$ [cf. (a)], and also when $n = 2$ because

$$\begin{vmatrix} 1 & x_1 \\ 1 & x_2 \end{vmatrix} = (x_2 - x_1).$$

Proceeding as in the hint, we get

$$V_n = \begin{vmatrix} 1 & 0 & \dots & 0 & 0 \\ 1 & (x_2 - x_1) & \dots & x_2^{n-3}(x_2 - x_1) & x_2^{n-2}(x_2 - x_1) \\ 1 & (x_3 - x_1) & \dots & x_3^{n-3}(x_3 - x_1) & x_3^{n-2}(x_3 - x_1) \\ \vdots & & & \dots & \\ 1 & (x_n - x_1) & \dots & x_n^{n-3}(x_n - x_1) & x_n^{n-2}(x_n - x_1) \end{vmatrix}$$

Expanding according to the first row, we get

$$V_n = (x_n - x_1)\cdots(x_2 - x_1)V_{n-1}$$

and by induction we suppose that $V_{n-1} = \prod_{1 \le i < j \le n}(x_j - x_i)$, so the result drops out.

6. *Find the determinant of the following matrices.*

(a) $\begin{pmatrix} 1 & 2 & 5 \\ 0 & 1 & 7 \\ 0 & 0 & 3 \end{pmatrix}$ (b) $\begin{pmatrix} -1 & 5 & 20 \\ 0 & 4 & 8 \\ 0 & 0 & 6 \end{pmatrix}$ (c) $\begin{pmatrix} 2 & -6 & 9 \\ 0 & 1 & 4 \\ 0 & 0 & 8 \end{pmatrix}$ (d) $\begin{pmatrix} -7 & 98 & 54 \\ 0 & 2 & 46 \\ 0 & 0 & -1 \end{pmatrix}$

(e) $\begin{pmatrix} 1 & 4 & 6 \\ 0 & 0 & 1 \\ 0 & 0 & 8 \end{pmatrix}$ (f) $\begin{pmatrix} 4 & 0 & 0 \\ -5 & 2 & 0 \\ 79 & 54 & 1 \end{pmatrix}$ (g) $\begin{pmatrix} 1 & 5 & 2 & 3 \\ 0 & 2 & 7 & 6 \\ 0 & 0 & 4 & 1 \\ 0 & 0 & 0 & 5 \end{pmatrix}$ (h) $\begin{pmatrix} -5 & 0 & 0 & 0 \\ 7 & 2 & 0 & 0 \\ -9 & 4 & 1 & 0 \\ 96 & 2 & 3 & 1 \end{pmatrix}$.

(i) Let A be a triangular $n \times n$ matrix, say a matrix such that all components below the diagonal are equal to 0. What is $D(A)$?

SOLUTION.
(a) 3 (b) −24 (c) 16 (d) 14 (e) 0 (f) 8 (g) 40

(h) −10.

(i) Expanding according to the first column, we get

$$\operatorname{Det}(A) = a_{11}\begin{vmatrix} a_{22} & a_{23} & \cdots & a_{2n} \\ 0 & & & \vdots \\ \vdots & & & a_{n-1n} \\ 0 & \cdots & 0 & a_{nn} \end{vmatrix};$$

therefore, we see that $\operatorname{Det}(A) = a_{11}a_{22}\cdots a_{nn}$.

7. *If $a(t)$, $b(t)$, $c(t)$, $d(t)$ are functions of t, one can form the determinant*

$$\begin{vmatrix} a(t) & b(t) \\ c(t) & d(t) \end{vmatrix}$$

just as with numbers. Write out in full the determinant

$$\begin{vmatrix} \sin t & \cos t \\ -\cos t & \sin t \end{vmatrix}.$$

SOLUTION. Using a trigonometric identity, we get

$$\begin{vmatrix} \sin t & \cos t \\ -\cos t & \sin t \end{vmatrix} = \sin^2 t + \cos^2 t = 1.$$

8. *Write out in full the determinant* $\begin{vmatrix} t+1 & t-1 \\ t & 2t+5 \end{vmatrix}$.

SOLUTION. $\begin{vmatrix} t+1 & t-1 \\ t & 2t+5 \end{vmatrix} = (t+1)(2t+5) - t(t-1) = t^2 + 8t + 5.$

9. *Let $f(t)$, $g(t)$ be two function having derivatives of all orders. Let $\varphi(t)$ be the function obtained by taking the determinant*

$$\varphi(t) = \begin{vmatrix} f(t) & g(t) \\ f'(t) & g'(t) \end{vmatrix}.$$

Show that

$$\varphi'(t) = \begin{vmatrix} f(t) & g(t) \\ f''(t) & g''(t) \end{vmatrix}.$$

SOLUTION. Since $\varphi = fg' - f'g$, we have

$$\varphi' = f'g' + fg'' - f''g - f'g' = fg'' - f''g = \begin{vmatrix} f & g \\ f'' & g'' \end{vmatrix}.$$

10. *Let* $A(t) = \begin{vmatrix} b_1(t) & c_1(t) \\ b_2(t) & c_2(t) \end{vmatrix}$ *be a* 2×2 *matrix of differentiable functions. Let* $B(t)$ *and* $C(t)$ *be its column vectors. Let* $\varphi(t) = \mathrm{Det}(A(t))$. *Show that*

$$\varphi'(t) = D(B'(t), C(t)) + D(B(t), C'(t)).$$

SOLUTION. Brute force shows that

$$\varphi' = b_1'c_2 + b_1c_2' - b_2'c_1 - b_2c_1',$$

and that

$$D(B', C) = b_1'c_2 - b_2'c_1, \quad D(B, C') = b_1c_2' - b_2c_1'.$$

So $\varphi'(t) = D(B'(t), C(t)) + D(B(t), C'(t))$.

11. *Let* $\alpha_1, \dots, \alpha_n$ *be distinct numbers* $\neq 0$. *Show that the functions*

$$e^{\alpha_1 t}, \dots, e^{\alpha_n t}$$

are linearly independent over the complex numbers. [Hint: Suppose we have a linear relation $c_1e^{\alpha_1 t} + \dots + c_ne^{\alpha_n t}$ *with constants* c_i *valid for all* t. *If not all* c_i *are 0, without loss of generality, we may assume that none of them is 0. Differentiate the above relation* $n-1$ *times. You get a system of linear equations. The determinant of its coefficients must be zero. (Why?) Get a contradiction from this.]*

SOLUTION. Differentiating $n-1$ times and setting $t = 0$ in each equation, we see that the system

$$\begin{cases} x_1 + \dots + x_n = 0 \\ x_1\alpha_1 + \dots + x_n\alpha_n = 0 \\ \vdots \\ x_1\alpha_1^{n-1} + \dots + x_n\alpha_1^{n-1} = 0 \end{cases}$$

has a nontrivial solution, namely, $(c_1, \dots, c_n)$. Therefore, the column vectors must be linearly dependent and hence the determinant

$$\begin{vmatrix} 1 & 1 & \dots & 1 \\ \alpha_1 & \alpha_2 & & \alpha_n \\ \vdots & \vdots & & \vdots \\ \alpha_1^{n-1} & \alpha_2^{n-1} & \dots & \alpha_n^{n-1} \end{vmatrix}$$

must be 0. But since $\alpha_1, \dots, \alpha_n$ are distinct, we see at once that the Vandermonde determinant is non-zero. We get a contradiction because the determinant of a matrix is equal to the determinant of its transpose.

VI, §4 Cramer's Rule

1. *Solve the following systems of linear equations.*

(a)
$$\begin{aligned} 3x + y - z &= 0 \\ x + y + z &= 0 \\ y - z &= 1 \end{aligned}$$

(b)
$$\begin{aligned} 2x - y + z &= 1 \\ x + 3y - 2z &= 0 \\ 4x - 3y + z &= 2 \end{aligned}$$

(c)
$$\begin{aligned} 4x + y + z + w &= 1 \\ x - y + 2z - 3w &= 0 \\ 2x + y + 3z + 5w &= 0 \\ x + y - z - w &= 2 \end{aligned}$$

(d)
$$\begin{aligned} x + 2y - 3z + 5w &= 0 \\ 2x + y - 4z - w &= 1 \\ x + y + z + w &= 0 \\ -x - y - z + w &= 4 \end{aligned}$$

SOLUTION.
(a) $x = \frac{-1}{3}$, $y = \frac{2}{3}$, $z = \frac{-1}{3}$

(b) $x = \frac{5}{12}$, $y = \frac{-1}{12}$, $z = \frac{1}{12}$

(c) $x = \frac{-5}{24}$, $y = \frac{97}{48}$, $z = \frac{1}{3}$, $w = \frac{-25}{48}$

(d) $x = \frac{11}{2}$, $y = \frac{-38}{5}$, $z = \frac{1}{10}$, $w = 2$.

VI, §5 Triangulation of a Matrix by Column Operations

1. *(a) Let $1 \le r, s \le n$ and $r \ne s$. Let J_{rs} be the $n \times n$ matrix whose rs-component is 1 and all other components are 0. Let $E_{rs} = I + J_{rs}$. Show that $D(E_{rs}) = 1$.*
(b) Let A be an $n \times n$ matrix. What is the effect of multiplying $E_{rs}A$? of multiplying AE_{rs}?

SOLUTION. (a) We see that adding the s row of I to the r row of I we get the matrix E_{rs}, so $D(E_{rs}) = D(I) = 1$.

s-column

$$E_{rs} = \begin{pmatrix} 1 & 0 & \cdots & 0 \\ 0 & & & \vdots \\ \vdots & & & \vdots \\ \cdots & 1 & 1 & \cdots \\ 0 & & \cdots & 0 \; 1 \end{pmatrix}$$

r-row

(b) When multiplying $E_{rs}A$, we add the s row of A to the r row of A, leaving the other rows of A unchanged. When multiplying AE_{rs}, we add the r column to the s column of A, leaving the other columns of A unchanged.

2. *In the proof of Theorem 5.3, we used the fact that if A is a triangular matrix, then the column vectors are linearly independent if and only if all the diagonal elements are* $\neq 0$. *Give the details of the proof of this fact.*

SOLUTION. (i) Suppose that all the diagonal elements are $\neq 0$. We want to solve

$$\begin{cases} x_1 b_{11} = 0 \\ x_1 b_{21} + x_2 b_{22} = 0 \\ \vdots \\ x_1 b_{nb1} + \ldots + x_n b_{nn} = 0 \end{cases}$$

From the first equation we get $x_1 = 0$. From the second equation we get $x_2 = 0$. Therefore we see that we must have $x_1 = x_2 = \ldots = x_n = 0$; so the column vectors are linearly independent.

(ii) Conversely, suppose that the column vectors are linearly independent and suppose that some diagonal element is 0. Then, since the row rank of the matrix is equal to the column rank, we see that if $b_{11} = 0$ or $b_{nn} = 0$, then we get a contradiction. Suppose that k is the smallest integer such that $b_{kk} = 0$ and $1 < k < n$. We contend that the system

$$\begin{cases} x_1 b_{11} = 0 \\ x_1 b_{21} + x_2 b_{22} = 0 \\ \vdots \\ x_1 b_{nb1} + \ldots + x_n b_{nn} = 0 \end{cases}$$

has a nontrivial solution. From (i) we see that we must have $x_1 = x_2 = \ldots = x_{k-1} = 0$. Therefore we are left with the system

$$\begin{cases} x_k b_{kk} = 0 \\ x_k b_{k+1k} + x_{k+1} b_{k+1k+1} = 0 \\ \vdots \\ x_k b_{nk} + \ldots + x_n b_{nn} = 0. \end{cases}$$

Deleting the first equation we see that the truncated system has one more unknown than it has of equations and therefore has a nontrivial solution. We then get a contradiction that proves the statement.

VI, §6 Permutations

In Exercise 1 we note τ_{ij}, the transposition that interchanges i and j. In each case, we can either write the permutation as a product of transposition or we can compute the determinants. We carry out both methods in (a) and (d).

1. *Determine the sign of the following permutations.*

(a) $\begin{bmatrix} 1 & 2 & 3 \\ 2 & 3 & 1 \end{bmatrix}$ *(b)* $\begin{bmatrix} 1 & 2 & 3 \\ 3 & 1 & 2 \end{bmatrix}$ *(c)* $\begin{bmatrix} 1 & 2 & 3 \\ 3 & 2 & 1 \end{bmatrix}$

(d) $\begin{bmatrix} 1 & 2 & 3 & 4 \\ 2 & 3 & 1 & 4 \end{bmatrix}$ *(e)* $\begin{bmatrix} 1 & 2 & 3 & 4 \\ 2 & 1 & 4 & 3 \end{bmatrix}$ *(f)* $\begin{bmatrix} 1 & 2 & 3 & 4 \\ 3 & 2 & 4 & 1 \end{bmatrix}$

(g) $\begin{bmatrix} 1 & 2 & 3 & 4 \\ 4 & 2 & 1 & 3 \end{bmatrix}$ *(h)* $\begin{bmatrix} 1 & 2 & 3 & 4 \\ 3 & 1 & 4 & 2 \end{bmatrix}$ *(i)* $\begin{bmatrix} 1 & 2 & 3 & 4 \\ 2 & 4 & 1 & 3 \end{bmatrix}$.

SOLUTION. (a) **Product of transpositions.** We see at once that

$$\tau_{12}\sigma = \tau_{23};$$

so $\sigma = \tau_{12}^{-1}\tau_{23}$, and therefore $\varepsilon(\sigma) = (-1)^2 = 1$.

Determinants. By definition,

$$\varepsilon(\sigma) = \frac{D(E^2, E^3, E^1)}{D(E^1, E^2, E^3)}.$$

But $D(E^1, E^2, E^3) = 1$, and expanding according to the first row we get

$$D(E^2, E^3, E^1) = \begin{vmatrix} 0 & 0 & 1 \\ 1 & 0 & 0 \\ 0 & 1 & 0 \end{vmatrix} = 1 \times \begin{vmatrix} 1 & 0 \\ 0 & 1 \end{vmatrix} = 1.$$

Thus $\varepsilon(\sigma) = 1$.

(b) $\varepsilon(\sigma) = 1$ because $\sigma = \tau_{13}^{-1}\tau_{23}$.

(c) $\varepsilon(\sigma) = -1$ because $\sigma = \tau_{13}$.

(d) **Product of transpositions.** We see at once that

$$\tau_{12}\sigma = \tau_{23};$$

so $\sigma = \tau_{12}^{-1}\tau_{23}$, and therefore $\varepsilon(\sigma) = (-1)^2 = 1$.

Determinants. By definition,

$$\varepsilon(\sigma) = \frac{D(E^2, E^3, E^1, E^4)}{D(E^1, E^2, E^3, E^4)}.$$

But $D(E^1, E^2, E^3, E^4) = 1$, and expanding according to the first row we get

$$D(E^2, E^3, E^1, E^4) = \begin{vmatrix} 0 & 0 & 1 & 0 \\ 1 & 0 & 0 & 0 \\ 0 & 1 & 0 & 0 \\ 0 & 0 & 0 & 1 \end{vmatrix} = 1 \times \begin{vmatrix} 1 & 0 & 0 \\ 0 & 1 & 0 \\ 0 & 0 & 1 \end{vmatrix} = 1.$$

Thus $\varepsilon(\sigma) = 1$.

(e) $\varepsilon(\sigma) = 1$ because $\sigma = \tau_{12}\tau_{34}$.

(f) $\varepsilon(\sigma) = 1$ because $\tau_{13}\sigma = \tau_{34}$.

(g) $\varepsilon(\sigma) = 1$ because $\tau_{14}\sigma = \tau_{34}$.

(h) $\varepsilon(\sigma) = -1$ because $\tau_{23}\tau_{13}\sigma = \tau_{34}$.

(i) $\varepsilon(\sigma) = -1$ because $\tau_{42}\tau_{12}\sigma = \tau_{34}$.

2. *In each of the cases of Exercise 1, write the inverse of the permutation.*

SOLUTION.

(a) $\sigma^{-1} = \begin{bmatrix} 1 & 2 & 3 \\ 3 & 1 & 2 \end{bmatrix}$ (b) $\sigma^{-1} = \begin{bmatrix} 1 & 2 & 3 \\ 2 & 3 & 1 \end{bmatrix}$ (c) $\sigma^{-1} = \begin{bmatrix} 1 & 2 & 3 \\ 3 & 2 & 1 \end{bmatrix}$

(d) $\sigma^{-1} = \begin{bmatrix} 1 & 2 & 3 & 4 \\ 3 & 1 & 2 & 4 \end{bmatrix}$ (e) $\sigma^{-1} = \begin{bmatrix} 1 & 2 & 3 & 4 \\ 2 & 1 & 4 & 3 \end{bmatrix}$ (f) $\sigma^{-1} = \begin{bmatrix} 1 & 2 & 3 & 4 \\ 4 & 2 & 1 & 3 \end{bmatrix}$

(g) $\sigma^{-1} = \begin{bmatrix} 1 & 2 & 3 & 4 \\ 3 & 2 & 4 & 1 \end{bmatrix}$ (h) $\sigma^{-1} = \begin{bmatrix} 1 & 2 & 3 & 4 \\ 2 & 4 & 1 & 3 \end{bmatrix}$ (i) $\sigma^{-1} = \begin{bmatrix} 1 & 2 & 3 & 4 \\ 3 & 1 & 4 & 2 \end{bmatrix}$.

3. *Show that the number of odd permutations of $\{1,\dots,n\}$ for $n \geq 2$ is equal to the number of even permutations. [Hint: Let τ be a transposition. Show that the map $\sigma \to \tau\sigma$ establishes an injective and surjective map between the even and odd permutations.]*

SOLUTION. Any permutation can be written as a product of transposition; so if σ is an even permutation, then we can write $\sigma = \tau_1\tau_2 \cdots \tau_s$, where s is even. Thus $\tau\sigma$ is odd, so $f\colon \sigma \to \tau\sigma$ is a map between the even and odd permutations of J_n.

Given an odd permutation $\sigma' = \tau_1\tau_2 \cdots \tau_p$ where p is odd, we see that $\sigma = \tau\tau_1\tau_2 \cdots \tau_p$ is even and that

$$f(\sigma) = \tau^2\tau_1\tau_2 \cdots \tau_p = \tau_1\tau_2 \cdots \tau_p = \sigma',$$

so f is surjective.

If $f(\sigma_1) = f(\sigma_2)$, then $\tau\sigma_1 = \tau\sigma_2$, so composing with τ we get $\sigma_1 = \sigma_2$, and therefore f is injective.

VI, §7 Expansion Formula and Uniqueness of Determinants

1. *Show that when $n = 3$, the expansion of Theorem 7.2 is the six-term expression given in §2.*

SOLUTION. The six permutations of J_3 into J_3 are given by

$\sigma_1 = \begin{bmatrix} 1 & 2 & 3 \\ 1 & 2 & 3 \end{bmatrix}$ $\sigma_2 = \begin{bmatrix} 1 & 2 & 3 \\ 1 & 3 & 2 \end{bmatrix}$ $\sigma_3 = \begin{bmatrix} 1 & 2 & 3 \\ 2 & 3 & 1 \end{bmatrix}$

$$\sigma_4 = \begin{bmatrix} 1 & 2 & 3 \\ 2 & 1 & 3 \end{bmatrix} \qquad \sigma_5 = \begin{bmatrix} 1 & 2 & 3 \\ 3 & 1 & 2 \end{bmatrix} \qquad \sigma_6 = \begin{bmatrix} 1 & 2 & 3 \\ 3 & 2 & 1 \end{bmatrix},$$

where $\varepsilon(\sigma_1) = \varepsilon(\sigma_3) = \varepsilon(\sigma_5) = 1$ and $\varepsilon(\sigma_2) = \varepsilon(\sigma_4) = \varepsilon(\sigma_6) = -1$. So the sum in Theorem 7.2 can be expressed as

$$D'(A) = \sum_{\sigma} \varepsilon(\sigma) a_{\sigma(1),1} \cdots a_{\sigma(3),3} = \sum_{i=1}^{6} \varepsilon(\sigma_i) a_{\sigma_i(1),1} \cdots a_{\sigma_i(3),3},$$

thus

$$D'(A) = a_{11}a_{22}a_{33} - a_{11}a_{32}a_{23} + a_{21}a_{32}a_{13} - a_{21}a_{12}a_{33} + a_{31}a_{12}a_{23} - a_{31}a_{22}a_{13}.$$

Expanding according to the first row, we see that the determinant of A is given by

$$D(A) = a_{11}(a_{22}a_{33} - a_{32}a_{23}) - a_{12}(a_{21}a_{33} - a_{31}a_{23}) + a_{13}(a_{21}a_{32} - a_{31}a_{22}).$$

Therefore $D'(A) = D(A)$.

2. *Go through the proof of Lemma 7.1 to verify that you did not use all the properties of the determinants in the proof. You used only the first two properties. Thus let F be any multilinear, alternating function. As in Lemma 7.1, let $A^j = \sum_{i=1}^{n} b_{ij} X^i$ for $j = 1, \ldots, n$. Then*

$$F(A^1, \ldots, A^n) = \sum_{\sigma} \varepsilon(\sigma) b_{\sigma(1),1} \cdots b_{\sigma(n),n} F(X^1, \ldots, X^n).$$

Why can you conclude that if B is the matrix (b_{ij}), then

$$F(A^1, \ldots, A^n) = D(B) F(X^1, \ldots, X^n)?$$

SOLUTION. In Lemma 7.1 we use the linearity property with respect to each column, and the fact that if two columns are equal, then the determinant is 0; so we used only properties 1 and 2, which are the properties of any alternating multilinear function. We also used the fact that

$$D(X^{\sigma(1)}, \ldots, X^{\sigma(n)}) = \varepsilon(\sigma) D(X^1, \ldots, X^n).$$

In order to apply Lemma 7.1 to any alternating multilinear function F, we must show that $F(X^{\sigma(1)}, \ldots, X^{\sigma(n)}) = \varepsilon(\sigma) F(X^1, \ldots, X^n)$. But this is obvious because any permutation is a product of transpositions, and each transposition changes the sign of F. Hence

$$F(A^1,\dots,A^n)=\sum_\sigma \varepsilon(\sigma)b_{\sigma(1),1}\cdots b_{\sigma(n),n}F(X^1,\dots,X^n).$$

Since $D(B)=\sum_\sigma \varepsilon(\sigma)b_{\sigma(1),1}\cdots b_{\sigma(n),n}$, we see that

$$F(A^1,\dots,A^n)=D(B)F(X^1,\dots,X^n).$$

3. *Let $F\colon \mathbf{R}^n\times\cdots\mathbf{R}^n\to\mathbf{R}$ be a function of n variables, each of which ranges over $\mathbf{R}^n$. Assume that F is linear in each variable, and that if $A^1,\dots,A^n\in\mathbf{R}^n$ and if there exists a pair of integers r, s with $1\le r,s\le n$ such that $r\ne s$ and $A^r=A^s$ then $F(A^1,\dots,A^n)=0$. Let B^i $(i=1,\dots,n)$ be vectors and c_{ij} numbers such that $A^j=\sum_{i=1}^n c_{ij}B^i$.*

(a) If $F(B^1,\dots,B^n)=-3$ and $\det(c_{ij})=5$, what is $F(A^1,\dots,A^n)$? Justify your answer by citing appropriate theorems, or proving it.
(b) If $F(E^1,\dots,E^n)=2$ (where $E^1,\dots,E^n$ are the standard unit vectors), and if $F(A^1,\dots,A^n)=10$, what is $D(A^1,\dots,A^n)$? Again give reasons for your answer.

SOLUTION. (a) The function F is multilinear and alternating, so we can apply the formula of Exercise 2, namely,

$$F(A^1,\dots,A^n)=\det(C)F(B^1,\dots,B^n),$$

where $C=(c_{ij})$. Hence $F(A^1,\dots,A^n)=-15$.

(b) Let $A^j=\sum_{i=1}^n a_{ij}E^j$. Then the matrix (a_{ij}) is the matrix whose columns are $A^1,\dots,A^n$; therefore, by Exercise 2 we get

$$F(A^1,\dots,A^n)=\det(a_{ij})F(E^1,\dots,E^n).$$

But $\det(a_{ij})=D(A^1,\dots,A^n)$, so $D(A^1,\dots,A^n)=5$.

VI, §8 Inverse of a Matrix

1. *Find the inverses of the matrices in Exercise 1, §3.*

SOLUTION.

(a) $\begin{pmatrix} \frac{-1}{5} & \frac{-1}{20} & \frac{7}{20} \\ \frac{1}{5} & \frac{3}{10} & \frac{-1}{10} \\ \frac{3}{5} & \frac{-1}{10} & \frac{-3}{10} \end{pmatrix}$ (b) $\begin{pmatrix} \frac{2}{5} & \frac{23}{5} & \frac{-11}{5} \\ \frac{1}{5} & \frac{19}{5} & \frac{-8}{5} \\ 0 & -2 & 1 \end{pmatrix}$ (c) $\begin{pmatrix} \frac{3}{4} & \frac{1}{2} & \frac{-9}{4} \\ \frac{1}{4} & \frac{1}{2} & \frac{-3}{4} \\ \frac{-1}{2} & -1 & \frac{5}{2} \end{pmatrix}$

(d) $\begin{pmatrix} 1 & \frac{-16}{5} & \frac{3}{5} \\ 0 & \frac{7}{5} & \frac{-1}{5} \\ 0 & \frac{-2}{5} & \frac{1}{5} \end{pmatrix}$ (e) $\begin{pmatrix} 0 & \frac{1}{4} & 0 \\ \frac{8}{19} & \frac{7}{38} & \frac{-3}{19} \\ \frac{-7}{19} & \frac{-17}{76} & \frac{5}{19} \end{pmatrix}$ (f) $\begin{pmatrix} \frac{17}{14} & \frac{-1}{2} & \frac{9}{14} \\ \frac{-11}{14} & \frac{1}{2} & \frac{-5}{14} \\ \frac{-13}{14} & \frac{1}{2} & \frac{-11}{14} \end{pmatrix}$.

2. *Using the fact that if A, B are two $n \times n$ matrices then*

$$\operatorname{Det}(AB) = \operatorname{Det}(A)\operatorname{Det}(B)$$

prove that a matrix A such that $\operatorname{Det}(A) = 0$ *does not have an inverse.*

SOLUTION. Suppose that A has an inverse. Then

$$1 = \operatorname{Det}(I) = \operatorname{Det}(AA^{-1}) = \operatorname{Det}(A)\operatorname{Det}(A^{-1}),$$

but $\operatorname{Det}(A) = 0$, so we get a contradiction.

3. *Write down explicitly the inverses of the* 2×2 *matrices:*

(a) $\begin{pmatrix} 3 & -1 \\ 1 & 4 \end{pmatrix}$ *(b)* $\begin{pmatrix} -2 & 1 \\ 1 & 1 \end{pmatrix}$ *(c)* $\begin{pmatrix} a & b \\ c & d \end{pmatrix}$

SOLUTION. (a) $\begin{pmatrix} \frac{4}{13} & \frac{1}{13} \\ \frac{-1}{13} & \frac{3}{13} \end{pmatrix}$ (b) $\begin{pmatrix} \frac{-1}{3} & \frac{1}{3} \\ \frac{1}{3} & \frac{2}{3} \end{pmatrix}$

(c) A direct computation or the formula given in the text shows that the inverse of $\begin{pmatrix} a & b \\ c & d \end{pmatrix}$ is

$$\frac{1}{ad - bc}\begin{pmatrix} d & -b \\ -c & a \end{pmatrix}.$$

4. *If A is an $n \times n$ matrix whose determinant is $\neq 0$, and B is a given vector in n-space, show that the system of linear equations $AX = B$ has a unique solution. If $B = O$, this solution is $X = O$.*

SOLUTION. The equation $AX = B$ is equivalent to

$$x_1 A^1 + \ldots + x_n A^n = B.$$

The determinant of A is non-zero, so the n column vectors $A^1, \ldots, A^n$ are linearly independent. The result follows because $\{A^1, \ldots, A^n\}$ is a basis for the n-space.

VI, §9 The Rank of a Matrix and Subdeterminants

Compute the ranks of the following matrices.

1. $\begin{pmatrix} 2 & 3 & 5 & 1 \\ 1 & -1 & 2 & 1 \end{pmatrix}$ *SOLUTION.* 2.

2. $\begin{pmatrix} 3 & 5 & 1 & 4 \\ 2 & -1 & 1 & 1 \\ 5 & 4 & 2 & 5 \end{pmatrix}$ *SOLUTION.* 2.

3. $\begin{pmatrix} 3 & 5 & 1 & 4 \\ 2 & -1 & 1 & 1 \\ 8 & 9 & 3 & 9 \end{pmatrix}$ *SOLUTION.* 2.

4. $\begin{pmatrix} 3 & 5 & 1 & 4 \\ 2 & -1 & 1 & 1 \\ 7 & 1 & 2 & 5 \end{pmatrix}$ *SOLUTION.* 3.

5. $\begin{pmatrix} -1 & 1 & 6 & 5 \\ 1 & 1 & 2 & 3 \\ -1 & 2 & 5 & 4 \\ 2 & 1 & 0 & 1 \end{pmatrix}$ *SOLUTION.* 4.

6. $\begin{pmatrix} 2 & 1 & 6 & 6 \\ 3 & 1 & 1 & -1 \\ 5 & 2 & 7 & 5 \\ -2 & 4 & 3 & 2 \end{pmatrix}$ *SOLUTION.* 3.

7. $\begin{pmatrix} 2 & 1 & 6 & 6 \\ 3 & 1 & 1 & -1 \\ 5 & 2 & 7 & 5 \\ 8 & 3 & 8 & 4 \end{pmatrix}$ *SOLUTION.* 2.

8. $\begin{pmatrix} 3 & 1 & 1 & -1 \\ -2 & 4 & 3 & 2 \\ -1 & 9 & 7 & 3 \\ 7 & 4 & 2 & 1 \end{pmatrix}$ *SOLUTION.* 3.

CHAPTER VII

Symmetric, Hermitian, and Unitary Operators

VII, §1 Symmetric Operators

1. *(a) A matrix A is called skew-symmetric if ${}^tA = -A$. Show that any matrix M can be expressed as a sum of a symmetric matrix and a skew-symmetric matrix one, and that the latter expression is uniquely determined. [Hint: Let $A = \frac{1}{2}(M + {}^tM)$.]*

(b) Prove that if A is skew-symmetric, then A^2 is symmetric.

(c) Let A be skew-symmetric. Show that $\operatorname{Det}(A)$ *is 0 if A is an $n \times n$ matrix and n is odd.*

SOLUTION. (a) See Exercise 14, §3 of Chapter III.

(b) Since

$$ {}^t(A^2) = {}^tA\,{}^tA = (-A)(-A) = A^2, $$

the matrix A^2 is symmetric.

(c) We know from Exercise 2, §2 of Chapter VI, that

$$ \operatorname{Det}(-A) = (-1)^n \operatorname{Det}(A) = -\operatorname{Det}(A). $$

But $\operatorname{Det}(A) = \operatorname{Det}({}^tA) = \operatorname{Det}(-A)$, so $\operatorname{Det}(A) = -\operatorname{Det}(A)$, whence we conclude that $\operatorname{Det}(A) = 0$.

2. *Let A be an invertible symmetric matrix. Show that A^{-1} is symmetric.*

SOLUTION. We know that $AA^{-1} = I$ and ${}^tI = I$, so

$$ I = {}^t(AA^{-1}) = {}^t(A^{-1})\,{}^tA = {}^t(A^{-1})A. $$

The inverse of a matrix is uniquely determined, so ${}^t(A^{-1}) = A^{-1}$; therefore, A^{-1} is symmetric.

3. *Show that a triangular symmetric matrix is diagonal.*

SOLUTION. We may assume without loss of generality that A is upper triangular. Then the transpose of A is lower triangular, and since, we must have $A = {}^tA$, we see at once that A must be diagonal.

4. *Show that the diagonal elements of a skew-symmetric matrix are equal to 0.*

SOLUTION. The condition ${}^tA = -A$ implies $a_{kk} = -a_{kk}$ for all $1 \le k \le n$, so the diagonal elements of A are zero.

5. *Let V be a finite dimensional vector space over the field K, with a non-degenerate scalar product. Let v_0, w_0 be elements of V. Let $A: V \to V$ be the linear map such that $A(v) = \langle v_0, v\rangle w_0$. Describe ${}^tA$.*

SOLUTION. We have

$$\langle A(v), w\rangle = \langle v_0, v\rangle\langle w_0, w\rangle = \langle v, \langle w_0, w\rangle v_0\rangle,$$

so ${}^tA(w) = \langle w_0, w\rangle v_0$.

6. *Let V be the vector space over $\mathbf{R}$ of polynomials of degree ≤ 5. Let the scalar product be defined as usual by*

$$\langle f, g\rangle = \int_0^1 f(t)\, dt.$$

Describe the transpose of the derivative D with respect to this scalar product.

SOLUTION. Consider the basis $\{1, t, t^2, t^3, t^4, t^5\}$ for the space of polynomials of degree ≤ 5. Then the matrix A associated with the given scalar product is

$$A = \begin{pmatrix} 1 & \frac{1}{2} & \frac{1}{3} & \frac{1}{4} & \frac{1}{5} & \frac{1}{6} \\ \frac{1}{2} & \frac{1}{3} & \frac{1}{4} & \frac{1}{5} & \frac{1}{6} & \frac{1}{7} \\ \frac{1}{3} & \frac{1}{4} & \frac{1}{5} & \frac{1}{6} & \frac{1}{7} & \frac{1}{8} \\ \frac{1}{4} & \frac{1}{5} & \frac{1}{6} & \frac{1}{7} & \frac{1}{8} & \frac{1}{9} \\ \frac{1}{5} & \frac{1}{6} & \frac{1}{7} & \frac{1}{8} & \frac{1}{9} & \frac{1}{10} \\ \frac{1}{6} & \frac{1}{7} & \frac{1}{8} & \frac{1}{9} & \frac{1}{10} & \frac{1}{11} \end{pmatrix}.$$

We then have $\langle X, Y\rangle = {}^tXAY$. The matrix associated with D is given by

$$D = \begin{pmatrix} 0 & 1 & 0 & 0 & 0 & 0 \\ 0 & 0 & 2 & 0 & 0 & 0 \\ 0 & 0 & 0 & 3 & 0 & 0 \\ 0 & 0 & 0 & 0 & 4 & 0 \\ 0 & 0 & 0 & 0 & 0 & 5 \\ 0 & 0 & 0 & 0 & 0 & 0 \end{pmatrix}.$$

Then we see that

$$\langle DX, Y\rangle = {}^t(DX)AY = {}^tX\,{}^tDAY = {}^tXA\left(A^{-1}\,{}^tDA\right)Y;$$

so the transpose of the differential operator D is described by the matrix $A^{-1}\,{}^tDA$ with respect to the chosen basis.

7. *Let V be a finite dimensional space over the field K, with a non-degenerate scalar product. Let A: $V \to V$ be a linear map. Show that the image of ${}^tA$ is the orthogonal space to the kernel of A.*

SOLUTION. First we show that $\operatorname{Im}({}^tA)^{\perp} = \operatorname{Ker}(A)$. If $x \in \operatorname{Ker}(A)$ and $w' \in \operatorname{Im}({}^tA)$, where ${}^tA(w) = w'$, then

$$\langle x, w'\rangle = \langle x, {}^tA(w)\rangle = \langle A(x), w\rangle = 0,$$

so $x \in \operatorname{Im}({}^tA)^{\perp}$. Conversely, suppose $x \in \operatorname{Im}({}^tA)^{\perp}$; then for all $v \in V$ we have

$$0 = \langle x, {}^tA(v)\rangle = \langle A(x), v\rangle.$$

But the scalar product is non-degenerate, hence $A(x) = O$, which proves the assertion. Then Exercise 7, §6 of Chapter V, implies

$$\operatorname{Im}({}^tA) = \operatorname{Ker}(A)^{\perp}.$$

8. *Let V be a finite dimensional space over **R**, with a positive definite scalar product. Let P: $V \to V$ be a linear map such that $PP = P$. Assume that ${}^tPP = P\,{}^tP$ show that $P = {}^tP$.*

SOLUTION. We have

$$\langle {}^tP(v), {}^tP(v)\rangle = \langle v, P\,{}^tP(v)\rangle = \langle v, {}^tPP(v)\rangle = \langle P(v), P(v)\rangle,$$

so we see that $\operatorname{Ker}(P) = \operatorname{Ker}({}^tP)$ and

$$\langle P(v) - {}^tP(v), P(v) - {}^tP(v)\rangle = 2[\langle P(v), P(v)\rangle - \langle P(v), {}^tP(v)\rangle].$$

Exercise 10, §4 of Chapter IV, implies that $V = \operatorname{Im}(P) \oplus \operatorname{Ker}(P)$, so we can write $v = P(x) + w$, where $w \in \operatorname{Ker}(P) = \operatorname{Ker}({}^tP)$. Then

$$\langle P(v), P(v)\rangle = \langle P(x), P(x)\rangle$$

and

$$\langle P(v), {}^tP(v)\rangle = \langle P(x), {}^tPP(x)\rangle = \langle P(x), P(x)\rangle.$$

The scalar product is positive definite, so $P(v) = {}^tP(v)$.

9. *A square $n \times n$ real symmetric matrix A is said to be* **positive definite** *if ${}^tXAX > 0$ for all $X \neq O$. If A, B are symmetric (of the same size) we define $A < B$ to mean that $B - A$ is positive definite. Show that if $A < B$ and $B < C$ then $A < C$.*

SOLUTION. We have

$${}^tX(C-A)X = {}^tX(C-B+B-A)X = {}^tX(C-B)X + {}^tX(B-A)X,$$

so the result drops out.

10. *Let V be a finite dimensional vector space over* **R** *with a positive definite scalar product $\langle\ ,\ \rangle$. An operator A of V is said to be* **semipositive** *if $\langle Av, v\rangle \geq 0$ for all $v \in V$, $v \neq O$. Suppose that $V = W + W^\perp$ is the direct sum of a subspace W and its orthogonal complement. Let P be the projection on W, and assume that $W \neq \{O\}$. Show that P is symmetric and semipositive.*

SOLUTION. **The operator *P* is symmetric.** For $i = 1, 2$ write $v_i = w_i + w_i^\perp$, where $w_i \in W$ and $w_i^\perp \in W^\perp$. Then

$$\langle P(v_1), v_2\rangle = \langle w_1, w_2 + w_2^\perp\rangle = \langle w_1, w_2\rangle$$

and

$$\langle v_1, P(v_2)\rangle = \langle w_1 + w_1^\perp, w_2\rangle = \langle w_1, w_2\rangle,$$

so P is symmetric.

The operator P is semipositive. If $v = w + w^{\perp}$, where $w \in W$ and $w^{\perp} \in W^{\perp}$, then

$$\langle P(v), v\rangle = \langle w, w + w^{\perp}\rangle = \langle w, w\rangle \geq 0.$$

11. *Let the notation be as in Exercise 10. Let c be a real number, and let A be the operator such that $Av = cw$ if we can write $v = w + w'$ with $w \in W$ and $w' \in W^{\perp}$. Show that A is symmetric.*

SOLUTION. For $i = 1, 2$ we write $v_i = w_i + w_i'$, where $w_i \in W$ and $w_i' \in W^{\perp}$. Then

$$\langle A(v_1), v_2\rangle = \langle cw_1, w_2 + w_2'\rangle = c\langle w_1, w_2\rangle$$

and

$$\langle v_1, A(v_2)\rangle = \langle w_1 + w_1', cw_2\rangle = c\langle w_1, w_2\rangle.$$

Thus A is symmetric.

12. *Let the notation be as in Exercise 10. Let P be again the projection on W. Show that there is a symmetric operator A such that $A^2 = I + P$.*

SOLUTION. If $v = w + w'$, where $w \in W$ and $w' \in W^{\perp}$, we define

$$A(v) = \sqrt{2}w + w'.$$

Then

$$A^2(v) = A\left(\sqrt{2}w + w'\right) = 2w + w' = I(v) + P(v).$$

With the notation being the one of Exercise 11, we see that

$$\langle A(v_1), v_2\rangle = \langle \sqrt{2}w_1 + w_1', w_2 + w_2'\rangle = \sqrt{2}\langle w_1, w_2\rangle + \langle w_1', w_2'\rangle$$

and

$$\langle v_1, A(v_2)\rangle = \langle w_1 + w_1', \sqrt{2}w_2 + w_2'\rangle = \sqrt{2}\langle w_1, w_2\rangle + \langle w_1', w_2'\rangle,$$

so A is symmetric.

13. *Let A be a real symmetric matrix. Show that there exists a real number c so that $A + cI$ is positive.*

SOLUTION. We have

$$'XAX = \sum_{i,j} a_{ij}x_ix_j \geq \sum_{i,j} -|a_{ij}||x_k|^2$$

for some k, so if we choose $c > \sum_{i,j} |a_{ij}|$, then $A + cI$ is positive because

$$'X(A+cI)X = 'XAX + c(x_1^2 + \ldots + x_n^2).$$

14. *Let V be a finite dimensional vector space over the field K, with a non-degenerate scalar product $\langle\, ,\, \rangle$. If A: $V \to V$ is a linear map such that*

$$\langle Av, Aw\rangle = \langle v, w\rangle$$

for all v, $w \in V$, show that $\mathrm{Det}(A) = \pm 1$.

SOLUTION. Fix w. Then for all v in V we have

$$\langle v, w\rangle = \langle Av, Aw\rangle = \langle v, 'AAw\rangle.$$

Thus

$$\langle v, w - 'AAw\rangle = 0,$$

so $'AA = I$ because the scalar product in non-degenerate. Therefore,

$$1 = \mathrm{Det}(I) = \mathrm{Det}('AA) = \mathrm{Det}('A)\mathrm{Det}(A),$$

but $\mathrm{Det}('A) = \mathrm{Det}(A)$, so we conclude that $\mathrm{Det}(A) = \pm 1$. In the general case we also have $A^*A = AA^* = I$ (where A^* is the transpose of the operator). If J represents the scalar product, then $\langle Av, w\rangle = \langle v, A^*w\rangle$ so that $A'J = JA'$ where A' is the transpose matrix of A. Hence

$$\mathrm{Det}(A') = \mathrm{Det}(A^*),$$

which implies that $\mathrm{Det}(A^*)^2 = 1$.

15. *Let A, B be symmetric matrices of the same size over the field K. Show that AB is symmetric if and only if $AB = BA$.*

SOLUTION. Since

$$'(AB) = 'B'A = BA,$$

we see that AB is symmetric if and only if $AB = BA$.

VII, §2 Hermitian Operators

1. *Let A be an invertible hermitian matrix. Show that A^{-1} is hermitian.*

SOLUTION. The matrix A^{-1} is hermitian because

$$^t(A^{-1}) = (^tA)^{-1} = (\bar{A})^{-1} = \overline{(A^{-1})}.$$

2. *Show that the analogue of Theorem 2.4 when V is a finite dimensional space over* **R** *is false. In other words, it may happen that Av is perpendicular to all $v \in V$ without A being the zero map!*

SOLUTION. A suitable rotation maps a vector into a vector orthogonal to it.

3. *Show that the analogue of Theorem 2.4 when V is a finite dimensional space over* **R** *is true if we assume in addition that A is symmetric.*

SOLUTION. The polarization identity is also true in a finite dimensional vector space over **R**, so

$$\langle Aw, v\rangle + \langle Av, w\rangle = 0$$

for all $v, w \in V$. But A is symmetric, so

$$2\langle Av, w\rangle = 0.$$

Conclude.

4. *Which of the following matrices are hermitian?*

$$(a)\begin{pmatrix} 2 & i \\ -i & 5 \end{pmatrix} \qquad (b)\begin{pmatrix} 1+i & 2 \\ 2 & 5i \end{pmatrix} \qquad (c)\begin{pmatrix} 1 & 1+i & 5 \\ 1-i & 2 & i \\ 5 & -i & 7 \end{pmatrix}.$$

SOLUTION. The matrices of (a) and (c) are hermitian because $^tA = \bar{A}$.

5. *Show that the diagonal elements of a hermitian matrix are real.*

SOLUTION. The diagonal elements verify $a_{kk} = \overline{a}_{kk}$, so they must be real.

6. *Show that a triangular hermitian matrix is diagonal.*

SOLUTION. The conjugate transpose of a lower (resp. upper) triangular matrix is an upper (resp. lower) triangular matrix. The result follows.

7. *Let A, B be hermitian matrices (of the same size). Show that $A+B$ is hermitian. If $AB = BA$, show that AB is hermitian.*

SOLUTION. The matrix $A+B$ is hermitian because

$$ {}^t(A+B) = {}^tA + {}^tB = \overline{A} + \overline{B} = \overline{A+B}. $$

Under the assumption that $AB = BA$, the matrix AB is hermitian because

$$ {}^t(AB) = {}^tB\,{}^tA = \overline{B}\overline{A} = \overline{BA} = \overline{AB}. $$

8. *Let V be a finite dimensional vector space over* **C**, *with a positive definite hermitian product. Let $A\colon V \to V$ be a hermitian operator. Show that $I+iA$ and $I-iA$ are invertible. [Hint: If $v \neq O$, show that $\|(I+iA)v\| \neq 0$.]*

SOLUTION. We contend that $\operatorname{Ker}(I+iA) = \{O\}$. We have

$$ \begin{aligned} \langle (I+iA)v, (I+iA)v\rangle &= \langle v, v\rangle + \langle v, iAv\rangle + \langle iAv, v\rangle + \langle iAv, iAv\rangle \\ &= \langle v, v\rangle - i\langle v, Av\rangle + i\langle v, Av\rangle + \langle Av, Av\rangle \\ &= \langle v, v\rangle + \langle Av, Av\rangle. \end{aligned} $$

Since the product is positive definite, our contention is proved. Theorem 3.3 of Chapter III guarantees that $I+iA$ is invertible. The same argument shows that $I-iA$ is invertible.

9. *Let A be a hermitian matrix. Show that ${}^tA$ and $\overline{A}$ are hermitian. If A is invertible, show that A^{-1} is hermitian.*

SOLUTION. The matrix ${}^tA$ is hermitian because

$$ {}^t({}^tA) = {}^t(\overline{A}) = \overline{{}^tA}. $$

The matrix $\overline{A}$ is hermitian because

$$ {}^t(\overline{A}) = \overline{({}^tA)} = A = \overline{\overline{A}}. $$

See Exercise 1 for the proof that A^{-1} is hermitian.

10. *Let V be a finite dimensional space over* $\mathbf{C}$, *with a positive definite hermitian form* $\langle\ ,\ \rangle$. *Let $A\colon V \to V$ be a linear map. Show that the following conditions are equivalent:*
(i) We have $AA^* = A^*A$.
(ii) For all $v \in V$, $\|Av\| = \|A^*v\|$ *(where* $\|v\| = \sqrt{\langle v, v\rangle}$*).*
(iii) We can write $A = B + iC$, *where B, C are hermitian, and* $BC = CB$.

SOLUTION.
(i) implies (ii). Indeed,

$$\begin{aligned}\|Av\|^2 &= \langle Av, Av\rangle = \langle v, A^*Av\rangle = \langle v, AA^*v\rangle = \overline{\langle AA^*v, v\rangle}\\ &= \overline{\langle A^*v, A^*v\rangle} = \langle A^*v, A^*v\rangle = \|A^*v\|^2.\end{aligned}$$

(ii) implies (i). The assumption implies

$$\begin{aligned} & \langle Av, Av\rangle = \langle A^*v, A^*v\rangle \\ \Rightarrow\quad & \langle A^*Av, v\rangle = \langle AA^*v, v\rangle \\ \Rightarrow\quad & \langle (A^*A - AA^*)v, v\rangle = 0,\end{aligned}$$

so Theorem 2.4 implies that A and A^* commute.

(iii) implies (i). Since $A^* = B^* - iC^* = B - iC$ and $BC = CB$, we have

$$AA^* = (B + iC)(B - iC) = B^2 + C^2$$

and

$$A^*A = (B - iC)(B + iC) = B^2 + C^2.$$

(i) implies (iii). Let $B = \frac{1}{2}(A + A^*)$ and $C = \frac{1}{2i}(A - A^*)$. Then we have $B + iC = A$ and

$$2\langle Bv, w\rangle = \langle Av, w\rangle + \overline{\langle w, A^*v\rangle} = \langle v, A^*w\rangle + \langle v, Aw\rangle = 2\langle v, Bv\rangle,$$

so B is hermitian. A similar argument shows that C is hermitian. Finally, B and C commute because A and A^* commute. Indeed,

$$4iBC = \left(A^2 - AA^* + A^*A - A^*A^*\right)$$

and

$$4iCB = (A^2 + AA^* - A^*A - A^*A^*).$$

11. *Let A be a non-zero hermitian matrix. Show that* $\operatorname{tr}(AA^*) > 0$.

SOLUTION. If $A = (a_{ij})$, $A^* = (a^*_{ij})$ and $AA^* = (c_{ij})$, then

$$c_{kk} = \sum_{j=1}^{n} a_{kj}a^*_{jk} = \sum_{j=1}^{n} a_{kj}\bar{a}_{kj} = \sum_{j=1}^{n} |a_{kj}|^2.$$

VII, §3 Unitary Operators

1. *(a) Let V be a finite dimensional space over* **R**, *with a positive definite scalar product. Let* $\{v_1, \dots, v_n\}$ *and* $\{w_1, \dots, w_n\}$ *be orthonormal bases. Let* $A\colon V \to V$ *be an operator of V such that* $Av_i = w_i$. *Show that A is real unitary.*
(b) State and prove the analogue result in the complex case.

SOLUTION. (a) If $x = \sum_i a_i v_i$ and $y = \sum_i b_i v_i$, then

$$\langle x, y \rangle = \sum_i a_i b_i$$

and

$$\langle Ax, Ay \rangle = \langle \sum_i a_i Av_i, \sum_i b_i Av_i \rangle = \sum_i a_i b_i,$$

whence A is real unitary.

(b) We prove Theorem 3.4 in the complex case. Using the notation of part (a) we see that

$$\langle x, y \rangle = \sum_i a_i \bar{b}_i,$$

and

$$\langle Ax, Ay \rangle = \langle \sum_i a_i Av_i, \sum_i b_i Av_i \rangle = \sum_i a_i \bar{b}_i,$$

so A is complex unitary.

2. *Let V be as in Exercise 1. Let $\{v_1,\ldots,v_n\}$ be an orthonormal basis of V. Let A be a unitary operator of V. Show that $\{Av_1,\ldots,Av_n\}$ is an orthonormal basis.*

SOLUTION. For all $i \neq j$ we have

$$\langle Av_i, Av_j\rangle = \langle v_i, v_j\rangle = 0,$$

and

$$\langle Av_i, Av_i\rangle = \langle v_i, v_i\rangle = 1,$$

so $\{Av_1,\ldots,Av_n\}$ is an orthonormal basis of V.

3. *Let A be real unitary matrix.*
(a) Show that ${}^tA$ is unitary.
(b) Show that A^{-1} exists and is unitary.
(c) If B is real unitary, show that AB is unitary, and that $B^{-1}AB$ is unitary.

SOLUTION. (a) The matrix ${}^tA$ is real unitary because

$${}^t({}^tA) = A = ({}^tA)^{-1}.$$

(b) Since $\langle Av, Av\rangle = \langle v, v\rangle$, we see that $\operatorname{Ker}(A) = \{O\}$; so the matrix A^{-1} exists and is real unitary because

$${}^t(A^{-1}) = ({}^tA)^{-1} = (A^{-1})^{-1}.$$

(c) The matrix AB is real unitary because

$${}^t(AB) = {}^tB\,{}^tA = B^{-1}A^{-1} = (AB)^{-1}.$$

The matrix $B^{-1}AB$ is real unitary because

$${}^t(B^{-1}AB) = {}^tB\,{}^tA\,{}^tB^{-1} = B^{-1}A^{-1}B = (B^{-1}AB)^{-1}.$$

4. *Let A be complex unitary matrix.*
(a) Show that ${}^tA$ is unitary.
(b) Show that A^{-1} exists and is unitary.
(c) If B is complex unitary, show that AB is unitary, and that $B^{-1}AB$ is unitary.

SOLUTION. (a) (b) and (c) The arguments are the same as in Exercise 3 except for the complex conjugate matrix. We work out (c),

$$^t\left(\overline{B^{-1}AB}\right) = {}^t\left(\overline{B}^{-1}\overline{A}\overline{B}\right) = {}^t\overline{B}\,{}^t\overline{A}\,{}^t\overline{B}^{-1} = B^{-1}A^{-1}B = \left(B^{-1}AB\right)^{-1}.$$

5. *(a) Let V be a finite dimensional space over* **R**, *with a positive definite scalar product, and let* $\{v_1, \ldots, v_n\} = B$ *and* $\{w_1, \ldots, w_n\} = B'$ *be orthonormal bases of V. Show that the matrix* $M^B_{B'}(\mathrm{id})$ *is real unitary. [Hint: Use* $\langle w_i, w_i \rangle = 1$ *and* $\langle w_i, w_j \rangle = 0$ *if* $i \neq j$, *as well as the expression* $w_i = \sum a_{ij} v_j$ *for some* $a_{ij} \in \mathbf{R}$.*]*
(b) Let $F\colon V \to V$ *be such that* $F(v_i) = w_i$ *for all i. Show that* $M^B_{B'}(F)$ *is unitary.*

SOLUTION. (a) We can write $v_i = \sum_{j=1}^{n} a_{ji} w_j$. Then $M^{\mathscr{B}}_{\mathscr{B}'}(\mathrm{id}) = (a_{ij}) = A$. We contend that ${}^tAA = I$. If ${}^tAA = (c_{ij})$, then

$$c_{ij} = {}^tA^i \cdot A^j = \sum_{k=1}^{n} a_{ki} a_{kj}.$$

But

$$\langle v_i, v_j \rangle = \left\langle \sum_{k=1}^{n} a_{ki} w_k, \sum_{p=1}^{n} a_{pj} w_p \right\rangle = \sum_{k=1}^{n} a_{ki} a_{kj} = c_{ij},$$

thereby proving our contention.

(b) We simply have

$$M^B_{B'}(F) = I.$$

6. *Show that the absolute value of the determinant of a real unitary matrix is equal to 1. Conclude that if A is real unitary, then* $\mathrm{Det}(A) = 1$ *or* -1.

SOLUTION. Since ${}^tAA = I$ and $\mathrm{Det}({}^tAA) = \mathrm{Det}({}^tA)\mathrm{Det}(A) = [\mathrm{Det}(A)]^2$, the assertion drops out.

7. *If A is a complex square matrix, show that* $\mathrm{Det}(\overline{A}) = \overline{\mathrm{Det}(A)}$. *Conclude that the absolute value of the determinant of a complex unitary matrix is equal to 1.*

SOLUTION. The fact that $\mathrm{Det}(\bar{A}) = \overline{\mathrm{Det}(A)}$ follows from the formula for the determinant and the properties of the conjugate of a complex number, namely, the conjugate of a sum is the sum of the conjugates and the conjugate of a product is the product of the conjugates. As a consequence we have

$$\mathrm{Det}({}^t\bar{A}) = \overline{\mathrm{Det}(A)},$$

so if ${}^t\bar{A}A = I$, then

$$1 = \overline{\mathrm{Det}(A)}\,\mathrm{Det}(A) = |\mathrm{Det}(A)|^2.$$

8. *Let A be a diagonal real unitary matrix. Show that the diagonal elements of A are equal to 1 or* -1.

SOLUTION. All of the diagonal elements are $\neq 0$, and the inverse matrix is given by

$$A^{-1} = \begin{pmatrix} a_1^{-1} & 0 & \dots & 0 \\ 0 & a_2^{-1} & & \vdots \\ \vdots & & & 0 \\ 0 & \dots & 0 & a_n^{-1} \end{pmatrix}.$$

But $A = {}^tA$ and ${}^tA = A^{-1}$ so $a_j = a_j^{-1}$. Therefore, the diagonal elements of A are equal to 1 or -1.

9. *Let A be a diagonal complex matrix. Show that each diagonal element has absolute value 1, and hence is of type* $e^{i\theta}$, *with* θ *real.*

SOLUTION. The same argument as in Exercise 8 shows that $\bar{a}_j = a_j^{-1}$, so $|a_j|^2 = 1$.

The following exercises describe various properties of real unitary maps of the plane $\mathbf{R}^2$.

10. *Let V be a 2-dimensional vector space over* $\mathbf{R}$ *with a positive definite scalar product, and let A be a real unitary map of V into itself. Let* $\{v_1, v_2\}$ *and* $\{w_1, w_2\}$ *be orthonormal bases of V such that* $Av_i = w_i$ *for* $i = 1, 2$. *Let a, b, c, d be real numbers such that*

$$w_1 = av_1 + bv_2$$
$$w_2 = cv_1 + dv_2$$

Show that $a^2+b^2=1$, $c^2+d^2=1$, $ac+bd=0$, $a^2=d^2$, *and* $c^2=b^2$.

SOLUTION. Since $\langle w_1, w_1\rangle = \langle w_2, w_2\rangle = 1$, we immediately have that $a^2+b^2=c^2+d^2=1$, and $\langle w_1, w_2\rangle = 0$ implies $ac+bd=0$. One can either see geometrically that $a^2=d^2$ and $c^2=b^2$ by noting that the vectors (a,b) and (c,d) are perpendicular and that they both have norm one,

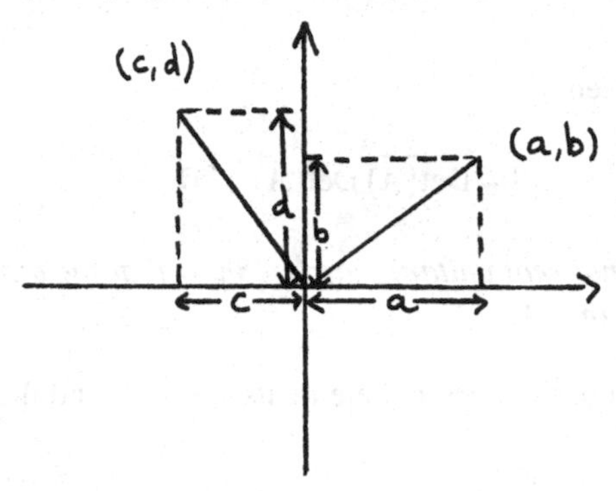

or one can do the algebra. For instance, if b and c are both non-zero then $a/b=-d/c$ and

$$b^2\left(1+\frac{a^2}{b^2}\right)=c^2\left(1+\frac{d^2}{c^2}\right)=c^2\left(1+\frac{a^2}{b^2}\right),$$

so $b^2=c^2$.

11. *Show that the determinant* $ad-bc$ *is equal to 1 or* -1. *(Show that its square is 1.)*

SOLUTION. Squaring the desired quantity we get

$$\begin{aligned}(ad-bc)^2 &= a^2d^2-2acbd+b^2c^2 = a^4+2a^2c^2+c^4\\ &= \left(a^2+c^2\right)^2 = \left(a^2+b^2\right)^2 = 1.\end{aligned}$$

12. *Define a rotation of V to be a real unitary map A of V whose determinant is 1. Show that the matrix of A relative to an orthogonal basis of V is of type*

$$\begin{pmatrix} a & -b \\ b & a \end{pmatrix}$$

for some real numbers a, b such that $a^2+b^2=1$. *Also prove the converse, that any linear map of V into itself represented by such a matrix on an or-*

thonormal basis is unitary, and has determinant 1. Using calculus, one can then conclude that there exist a number θ such that $a = \cos\theta$ and $b = \sin\theta$.

SOLUTION. Let $\{v_1, v_2\}$ be an orthogonal basis for V. Let $w_i = Av_i$ and

$$\begin{aligned} w_1 &= av_1 + bv_2 \\ w_2 &= cv_1 + dv_2. \end{aligned}$$

The matrix representing V in the chosen basis is

$$\begin{pmatrix} a & c \\ b & d \end{pmatrix}.$$

Then, since $\langle Av_i, Av_i\rangle = \langle v_i, v_i\rangle$, we have

$$\begin{aligned} (a^2-1)\langle v_1, v_1\rangle + b^2\langle v_2, v_2\rangle &= 0 \\ c^2\langle v_1, v_1\rangle + (d^2-1)\langle v_2, v_2\rangle &= 0. \end{aligned} \qquad (*)$$

But $dw_1 - bw_2 = (ad-bc)v_1 = v_1$, so

$$\langle v_1, v_1\rangle = \langle A(dv_1 - bv_2), A(dv_1 - bv_2)\rangle = d^2\langle v_1, v_1\rangle + b^2\langle v_2, v_2\rangle,$$

thus $(*)$ implies $a^2 = d^2$ and $b^2 = c^2$. Moreover,

$$0 = \langle v_1, v_2\rangle = \langle Av_1, Av_2\rangle = ac\langle v_1, v_1\rangle + bd\langle v_2, v_2\rangle,$$

so ac and bd are of opposite signs and therefore the matrix of A has the desired form.

Conversely, suppose that A is of the form described in the problem. Let $\{v_1, v_2\}$ be the orthonormal basis, and let $v = cv_1 + dv_2$. Then

$$Av = c(av_1 + bv_2) + d(-bv_1 + av_2) = (ca - bd)v_1 + (cb + da)v_2$$

so that

$$\langle Av, Av\rangle = (ca-bd)^2 + (cb+da)^2 = c^2 + d^2 = \langle v, v\rangle.$$

Hence A is unitary.

13. *Show that there exists a complex unitary matrix U such that if*

$$A = \begin{pmatrix} \cos\theta & -\sin\theta \\ \sin\theta & \cos\theta \end{pmatrix} \quad \textit{and} \quad B = \begin{pmatrix} e^{i\theta} & 0 \\ 0 & e^{-i\theta} \end{pmatrix}$$

then $U^{-1}AU = B$.

SOLUTION. Let $U_0 = \begin{pmatrix} \frac{1}{\sqrt{2}} & \frac{i}{\sqrt{2}} \\ \frac{i}{\sqrt{2}} & \frac{1}{\sqrt{2}} \end{pmatrix}$. Then $\det U_0 = 1$ and $U_0^{-1} = \begin{pmatrix} \frac{1}{\sqrt{2}} & \frac{-i}{\sqrt{2}} \\ \frac{-i}{\sqrt{2}} & \frac{1}{\sqrt{2}} \end{pmatrix} = {}^t\overline{U}_0$ so U_0 is complex unitary and one verifies that

$$U_0^{-1}BU_0 = A.$$

Let $U = U_0^{-1}$ and conclude.

14. *Let $V = \mathbf{C}$ be viewed as a vector space of dimension 2 over $\mathbf{R}$. Let $\alpha \in \mathbf{C}$, and let $L_\alpha : \mathbf{C} \to \mathbf{C}$ be the map $z \mapsto \alpha z$. Show that L_α is an $\mathbf{R}$-linear map of V into itself. For which complex numbers α is L_α a unitary map with respect to the scalar product $\langle z, w\rangle = \operatorname{Re}(z\overline{w})$? What is the matrix of L_α with respect to the basis $\{1, i\}$ of $\mathbf{C}$ over $\mathbf{R}$?*

SOLUTION. We have

$$L_\alpha(z_1 + z_2) = \alpha(z_1 + z_2) = \alpha z_1 + \alpha z_2 = L_\alpha(z_1) + L_\alpha(z_2)$$

and

$$L_\alpha(cz) = \alpha(cz) = c\alpha z = cL_\alpha(z).$$

We contend that L_α is unitary if and only if $|\alpha| = 1$. This result is obvious form the fact that

$$\langle \alpha z, \alpha w\rangle = \operatorname{Re}(\alpha z\overline{\alpha}\overline{w}) = |\alpha|^2 \operatorname{Re}(z\overline{w}) = |\alpha|^2\langle z, w\rangle.$$

Let $\alpha = a + ib$. Then

$$L_\alpha(1) = a + ib \quad\text{and}\quad L_\alpha(i) = ia - b,$$

so the matrix of L_α with respect to the basis $\{1, i\}$ is

$$\begin{pmatrix} a & -b \\ b & a \end{pmatrix}.$$

CHAPTER VIII

Eigenvectors and Eigenvalues

VIII, §1 Eigenvectors and Eigenvalues

1. *Let $a \in K$ and $a \neq 0$. Prove that the eigenvectors of the matrix*

$$\begin{pmatrix} 1 & a \\ 0 & 1 \end{pmatrix}$$

generate a 1-dimensional space, and give a basis for this space.

SOLUTION. The equation

$$\begin{pmatrix} 1 & a \\ 0 & 1 \end{pmatrix}\begin{pmatrix} x \\ y \end{pmatrix} = \lambda\begin{pmatrix} x \\ y \end{pmatrix}$$

is equivalent to the system

$$\begin{cases} x + ay = \lambda x \\ \quad\;\; y = \lambda y \end{cases}$$

If $y \neq 0$, then $\lambda = 1$ but we get a contradiction with the first equation, so assume that $y = 0$. Then we are left with $x = \lambda x$. So the eigenvectors of the matrix generate the 1-dimensional space $V = \{(x, 0),\ x \in K\}$. A basis for this space is given by the column vector $\begin{pmatrix} 1 \\ 0 \end{pmatrix}$.

2. *Prove that the eigenvectors of the matrix* $\begin{pmatrix} 2 & 0 \\ 0 & 2 \end{pmatrix}$ *generate a 2-dimensional space and give a basis for this space. What are the eigenvalues of this matrix?*

SOLUTION. We solve

$$\begin{pmatrix} 2 & 0 \\ 0 & 2 \end{pmatrix}\begin{pmatrix} x \\ y \end{pmatrix} = \lambda\begin{pmatrix} x \\ y \end{pmatrix}.$$

So we have

$$\begin{cases} 2x = \lambda x \\ 2y = \lambda y \end{cases}$$

therefore, the only eigenvalue of the matrix is $\lambda = 2$. The vectors $(1,0)$ and $(0,1)$ are eigenvectors of the matrix, and they are linearly independent.

3. *Let A be a diagonal matrix with diagonal elements $a_{11}, \ldots, a_{nn}$. What is the dimension of the space generated by the eigenvectors of A? Exhibit a basis for this space, and give the eigenvalues.*

SOLUTION. Since $AE^j = a_{jj}E^j$, the eigenvectors generate an n-dimensional space, and the eigenvalue of the eigenvector E^j is a_{jj}.

4. *Let $A = (a_{ij})$ be an $n \times n$ matrix such that for each $i = 1, \ldots, n$ we have $\sum_{j=1}^{n} a_{ij} = 0$. Show that 0 is an eigenvalue of A.*

SOLUTION. We have

$$\begin{pmatrix} a_{11} & \cdots & a_{1n} \\ \vdots & & \vdots \\ a_{n1} & \cdots & a_{nn} \end{pmatrix}\begin{pmatrix} 1 \\ \vdots \\ 1 \end{pmatrix} = \begin{pmatrix} 0 \\ \vdots \\ 0 \end{pmatrix}.$$

So if X_1 is the column vector with all ones, we have $AX_1 = 0X_1$; so 0 is an eigenvalue of A.

5. *(a) Show that if $\theta \in \mathbf{R}$, then the matrix*

$$A = \begin{pmatrix} \cos\theta & \sin\theta \\ \sin\theta & -\cos\theta \end{pmatrix}$$

always has an eigenvector in $\mathbf{R}^2$, *and in fact that there exists a vector* v_1 *such that* $Av_1 = v_1$. *[Hint: Let the first component of* v_1 *be*

$$x = \frac{\sin\theta}{1-\cos\theta}$$

if $\cos\theta \neq 1$. *Then solve for y. What if* $\cos\theta = 1$*?]*
(b) Let v_2 *be a vector of* $\mathbf{R}^2$ *perpendicular to the vector* v_1 *found in (a). Show that* $Av_2 = -v_2$. *Define this to mean that A is a reflection.*

SOLUTION. (a) **Case 1.** If $\cos\theta \neq 1$, consider the equation

$$\begin{pmatrix} \cos\theta & \sin\theta \\ \sin\theta & -\cos\theta \end{pmatrix}\begin{pmatrix} \dfrac{\sin\theta}{1-\cos\theta} \\ y \end{pmatrix} = \begin{pmatrix} \dfrac{\sin\theta}{1-\cos\theta} \\ y \end{pmatrix},$$

or, equivalently,

$$\begin{cases} \dfrac{\cos\theta\sin\theta}{1-\cos\theta} + y\sin\theta = \dfrac{\sin\theta}{1-\cos\theta} \\ \dfrac{\sin^2\theta}{1-\cos\theta} - y\cos\theta = y. \end{cases}$$

Solving for y we see from the second equation that $y = 1$. Conversely, one verifies at once that if

$$v_1 = \begin{pmatrix} \dfrac{\sin\theta}{1-\cos\theta} \\ 1 \end{pmatrix},$$

then $Av_1 = v_1$.

Case 2. If $\cos\theta = 1$, then $\sin\theta = 0$; so $v_1 = \begin{pmatrix} 1 \\ 0 \end{pmatrix}$ solves the equation $Av = v$.

(b) **Case 1.** If $\cos\theta \neq 1$, then a vector perpendicular to v_1 is

$$w = \begin{pmatrix} -1 \\ \dfrac{\sin\theta}{1-\cos\theta} \end{pmatrix}.$$

Then the first component of Aw is

$$-\cos\theta + \frac{\sin^2\theta}{1-\cos\theta} = \frac{1-\cos\theta}{1-\cos\theta} = 1,$$

and the second component of Aw is

$$-\sin\theta - \frac{\cos\theta\sin\theta}{1-\cos\theta} = -\frac{\sin\theta}{1-\cos\theta},$$

so $Aw = -w$, and therefore any multiple v_2 of w verifies $Av_2 = -v_2$.

Case 2. If $\cos\theta = 1$, then

$$A\begin{pmatrix}0\\1\end{pmatrix} = -\begin{pmatrix}0\\1\end{pmatrix},$$

so any multiple v_2 of $\begin{pmatrix}0\\1\end{pmatrix}$ verifies $Av_2 = -v_2$.

6. *Let*

$$R(\theta) = \begin{pmatrix}\cos\theta & -\sin\theta\\ \sin\theta & \cos\theta\end{pmatrix}$$

be the matrix of a rotation. Show that $R(\theta)$ does not have any real eigenvalues unless $R(\theta) = \pm I$. [It will be easier to do this exercise after you have read the next section.]

SOLUTION. We set the characteristic polynomial $P_{R(\theta)}$ equal to 0, namely,

$$\mathrm{Det}(\lambda I - R(\theta)) = 0.$$

We find

$$0 = (\lambda - \cos\theta)^2 + \sin^2\theta = \lambda^2 - 2\lambda\cos\theta + 1.$$

The discriminant of this equation is $4(\cos^2\theta - 1)$, so the matrix $R(\theta)$ has a real eigenvalue (1 or -1) if and only if $\cos\theta = \pm 1$, which is equivalent to $R(\theta) = \pm I$.

7. *Let V be a finite dimensional vector space. Let A, B be linear maps of V into itself. Assume that $AB = BA$. Show that if v is an eigenvector of A, with eigenvalue λ, then Bv is an eigenvector of A, with eigenvalue λ also if $Bv \neq O$.*

SOLUTION. We have

$$A(Bv) = B(Av) = B(\lambda v) = \lambda Bv.$$

VIII, §2 The Characteristic Polynomial

1. *Let A be a diagonal matrix with diagonal elements* $a_1, \ldots, a_n$.
(a) What is the characteristic polynomial of A?
(b) What are its eigenvalues?

SOLUTION. (a) The characteristic polynomial of A is

$$P_A(t) = (t - a_1)(t - a_2) \cdots (t - a_n).$$

(b) The eigenvalues are $a_1, a_2, \ldots, a_n$.

2. *Let A be a triangular matrix*

$$\begin{pmatrix} a_{11} & 0 & \ldots & 0 \\ a_{21} & a_{22} & \ldots & 0 \\ \vdots & \vdots & & \vdots \\ a_{n1} & a_{n2} & \ldots & a_{nn} \end{pmatrix}.$$

What is the characteristic polynomial of A, and what are its eigenvalues?

SOLUTION. Expanding according to the first row in each subdeterminant we find that the characteristic polynomial of A is

$$P_A(t) = (t - a_{11})(t - a_{22}) \cdots (t - a_{nn}),$$

so the eigenvalues are $a_{11}, a_{22}, \ldots, a_{nn}$.

In the following two exercises, find the characteristic polynomial, eigenvalues, and bases for the eigenspaces of the following matrices.

3. *(a)* $\begin{pmatrix} 1 & 2 \\ 3 & 2 \end{pmatrix}$ *(b)* $\begin{pmatrix} 3 & 2 \\ -1 & 0 \end{pmatrix}$ *(c)* $\begin{pmatrix} -2 & -7 \\ 1 & 2 \end{pmatrix}$ *(d)* $\begin{pmatrix} 1 & 4 \\ 2 & 3 \end{pmatrix}$.

SOLUTION.
(a) Since

$$P_A(t) = (t-1)(t-2) - 6 = t^2 - 3t - 4,$$

the eigenvalues are 4 and -1. We solve

$$\begin{cases} x+2y=\lambda x \\ 3x+2y=\lambda y. \end{cases}$$

Setting $\lambda=-1$ and fixing $y=1$, we find that $X_1=\begin{pmatrix}-1\\1\end{pmatrix}$ is a basis for V_{-1}.

Setting $\lambda=4$ and fixing $y=\frac{1}{2}$, we find that $X_2=\begin{pmatrix}\frac{2}{3}\\1\end{pmatrix}$ is a basis for V_4.

(b) Since

$$P_A(t)=t(t-3)+2=t^2-3t+2,$$

the eigenvalues are 1 and 2. We solve

$$\begin{cases} 3x+2y=\lambda x \\ -x=\lambda y. \end{cases}$$

Setting $\lambda=1$ and fixing $y=1$, we find that $X_1=\begin{pmatrix}-1\\1\end{pmatrix}$ is a basis for V_1.

Setting $\lambda=2$ and fixing $y=\frac{1}{2}$, we find that $X_2=\begin{pmatrix}-1\\\frac{1}{2}\end{pmatrix}$ is a basis for V_2.

(c) Since

$$P_A(t)=(t+2)(t-2)+7=t^2+3,$$

the eigenvalues are $i\sqrt{3}$ and $-i\sqrt{3}$. We solve

$$\begin{cases} -2x-7y=\lambda x \\ x+2y=\lambda y. \end{cases}$$

Setting $\lambda=i\sqrt{3}$ and fixing $y=1$, we find that $X_1=\begin{pmatrix}-2+i\sqrt{3}\\1\end{pmatrix}$ is a basis for $V_{i\sqrt{3}}$.

Setting $\lambda=-i\sqrt{3}$ and fixing $y=1$, we find that $X_2=\begin{pmatrix}-2-i\sqrt{3}\\1\end{pmatrix}$ is a basis for $V_{-i\sqrt{3}}$.

(d) Since

$$P_A(t) = (t-1)(t-3) - 8 = t^2 - 4t - 5,$$

the eigenvalues are 5 and -1. We solve

$$\begin{cases} x + 4y = \lambda x \\ 2x + 3y = \lambda y. \end{cases}$$

Setting $\lambda = 5$ and fixing $y = 1$, we find that $X_1 = \begin{pmatrix} 1 \\ 1 \end{pmatrix}$ is a basis for V_5.

Setting $\lambda = -1$ and fixing $y = 1$, we find that $X_2 = \begin{pmatrix} -2 \\ 1 \end{pmatrix}$ is a basis for V_{-1}.

4. *(a)* $\begin{pmatrix} 4 & 0 & 1 \\ -2 & 1 & 0 \\ -2 & 0 & 1 \end{pmatrix}$ *(b)* $\begin{pmatrix} 1 & -3 & 3 \\ 3 & -5 & 3 \\ 6 & -6 & 4 \end{pmatrix}$ *(c)* $\begin{pmatrix} 3 & 1 & 1 \\ 2 & 4 & 2 \\ 1 & 1 & 3 \end{pmatrix}$

(d) $\begin{pmatrix} 1 & 2 & 2 \\ 1 & 2 & -1 \\ -1 & 1 & 4 \end{pmatrix}$.

SOLUTION.
(a) The characteristic polynomial is given by

$$P_A(t) = (t-4)(t-1)^2 + 2(t-1) = (t-1)\left[t^2 - 5t + 6\right],$$

so the eigenvalues are 1, 2, and 3. We have

$$\begin{cases} 4x + z = \lambda x \\ -2x + y = \lambda y \\ -2x + z = \lambda z. \end{cases}$$

Setting $\lambda = 1$, we find that $X_1 = \begin{pmatrix} 0 \\ 1 \\ 0 \end{pmatrix}$ is a basis for V_1.

Setting $\lambda = 2$, we find that $X_2 = \begin{pmatrix} 1 \\ -2 \\ -2 \end{pmatrix}$ is a basis for V_2.

Setting $\lambda = 3$, we find that $X_3 = \begin{pmatrix} 1 \\ -1 \\ -1 \end{pmatrix}$ is a basis for V_3.

(b) The characteristic polynomial is

$$P_A(t) = t^3 - 12t - 16 = (t+2)^2(t-4),$$

so the eigenvalues are -2 and 4. The equation $AX = \lambda X$ is equivalent to

$$\begin{cases} x - 3y + 3z = \lambda x \\ 3x - 5y + 3z = \lambda y \\ 6x - 6y + 4z = \lambda z. \end{cases}$$

When $\lambda = -2$ the system is equivalent to the equation $x - y + z$, so a basis for V_{-2} is given by the two column vectors

$$\begin{pmatrix} 1 \\ 0 \\ -1 \end{pmatrix} \text{ and } \begin{pmatrix} 0 \\ 1 \\ 1 \end{pmatrix}.$$

When $\lambda = 4$ we see that a basis for V_4 is given by $\begin{pmatrix} 1 \\ 1 \\ 2 \end{pmatrix}$.

(c) The characteristic polynomial is

$$P_A(t) = t^3 - 10t^2 + 28t - 24 = (t-2)^2(t-6),$$

so the eigenvalues are 2 and 6. The equation $AX = \lambda X$ is equivalent to

$$\begin{cases} 3x + y + z = \lambda x \\ 2x + 4y + 2z = \lambda y \\ x + y + 3z = \lambda z. \end{cases}$$

When $\lambda = 2$ the above system is equivalent to $x + y + z = 0$, so a basis for V_2 is given by the two vectors

$$\begin{pmatrix}1\\0\\-1\end{pmatrix} \text{ and } \begin{pmatrix}0\\1\\-1\end{pmatrix}.$$

When $\lambda = 6$ we find that $x = z$ and $y = 2x$, so a basis for V_6 is $\begin{pmatrix}1\\2\\1\end{pmatrix}$.

(d) The characteristic polynomial is

$$P_A(t) = t^3 - 7t^2 + 15t - 9 = (t-1)(t-3)^2,$$

so the eigenvalues are 1 and 3. The equation $AX = \lambda X$ is equivalent to

$$\begin{cases} x + 2y + 2z = \lambda x \\ x + 2y - z = \lambda y \\ -x + y + 4z = \lambda z \end{cases}$$

When $\lambda = 3$ the above system is equivalent to $-x + y + z = 0$, so a basis for V_3 is given by the two vectors

$$\begin{pmatrix}1\\0\\1\end{pmatrix} \text{ and } \begin{pmatrix}0\\1\\-1\end{pmatrix}.$$

When $\lambda = 1$ the system implies that $y = -z$ and $x = 2z$, so a basis for V_1 is
given by the vector

$$\begin{pmatrix}2\\-1\\1\end{pmatrix}.$$

5. *Find the eigenvalues and eigenvectors of the following matrices. Show that the eigenvectors form a 1-dimensional space.*

(a) $\begin{pmatrix}2 & -1\\1 & 0\end{pmatrix}$ *(b)* $\begin{pmatrix}1 & 1\\0 & 1\end{pmatrix}$ *(c)* $\begin{pmatrix}2 & 0\\1 & 2\end{pmatrix}$ *(d)* $\begin{pmatrix}2 & -3\\1 & -1\end{pmatrix}$.

SOLUTION.
(a) The characteristic polynomial is

$$P_A(t) = (t-1)^2,$$

so the only eigenvalue is 1. The equation $AX = X$ is equivalent to

$$\begin{cases} 2x - y = x \\ x = y. \end{cases}$$

so the space generated by the eigenvectors is $V = \{(x, y) \text{ such that } x = y\}$. A basis for this space is $\begin{pmatrix} 1 \\ 1 \end{pmatrix}$, so V has dimension 1.

(b) The characteristic polynomial is

$$P_A(t) = (t-1)^2,$$

so the only eigenvalue is 1. The equation $AX = X$ is equivalent to

$$\begin{cases} x + y = x \\ y = y, \end{cases}$$

so the space generated by the eigenvectors is the space generated by $\begin{pmatrix} 1 \\ 0 \end{pmatrix}$.

(c) The characteristic polynomial is

$$P_A(t) = (t-2)^2,$$

so the only eigenvalue is 2. The equation $AX = 2X$ is equivalent to

$$\begin{cases} 2x = 2x \\ x + 2y = 2y, \end{cases}$$

so the space generated by the eigenvectors is the space generated by $\begin{pmatrix} 0 \\ 1 \end{pmatrix}$.

(d) The characteristic polynomial is

$$P_A(t) = t^2 - t + 1,$$

so the eigenvalues are

$$\lambda_1 = \frac{1+i\sqrt{3}}{2} \quad \text{and} \quad \lambda_2 = \frac{1-i\sqrt{3}}{2}.$$

The equation $AX = \lambda X$ is equivalent to

$$\begin{cases} 2x - 3y = \lambda x \\ x - y = \lambda y \end{cases} \Leftrightarrow \begin{cases} -3y = (\lambda - 2)x \\ x = (\lambda + 1)y. \end{cases}$$

Now we let $y = 1$ and find that two linearly independent eigenvectors are

$$X(\lambda_1) = \begin{pmatrix} \lambda_1 + 1 \\ 1 \end{pmatrix} \quad \text{and} \quad X(\lambda_2) = \begin{pmatrix} \lambda_2 + 1 \\ 1 \end{pmatrix}.$$

6. *Find the eigenvalues and eigenvectors of the following matrices. Show that the eigenvectors form a 1-dimensional space.*

$$(a) \begin{pmatrix} 1 & 1 & 1 \\ 0 & 1 & 1 \\ 0 & 0 & 1 \end{pmatrix} \qquad (b) \begin{pmatrix} 1 & 1 & 0 \\ 0 & 1 & 1 \\ 0 & 0 & 1 \end{pmatrix}.$$

SOLUTION.
(a) The characteristic polynomial is

$$P_A(t) = (t-1)^3,$$

so the only eigenvalue is 1. The equation $AX = X$ is equivalent to

$$\begin{cases} x + y + z = x \\ y + z = y \\ z = z. \end{cases}$$

We see that $y = z = 0$, so a basis for the eigenspace is $\begin{pmatrix} 1 \\ 0 \\ 0 \end{pmatrix}$.

(b) The characteristic polynomial is

$$P_A(t) = (t-1)^3,$$

so the only eigenvalue is 1. The equation $AX = X$ is equivalent to

$$\begin{cases} x+y=x \\ y+z=y \\ z=z. \end{cases}$$

so $y=z=0$, and therefore a basis for the eigenspace is $\begin{pmatrix}1\\0\\0\end{pmatrix}$.

7. *Find the eigenvalues and a basis for the eigenspaces of the following matrices.*

$$(a)\begin{pmatrix}0&1&0&0\\0&0&1&0\\0&0&0&1\\1&0&0&0\end{pmatrix} \qquad (b)\begin{pmatrix}-1&0&1\\-1&3&0\\-4&13&-1\end{pmatrix}.$$

SOLUTION.
(a) The characteristic polynomial is

$$P_A(t)=t^4-1,$$

so the eigenvalues of the matrix are $\lambda_1=1$, $\lambda_2=-1$, $\lambda_3=i$, and $\lambda_4=-i$. The following eigenvectors are a basis for the associated eigenspace:

$$X(\lambda_1)=\begin{pmatrix}1\\1\\1\\1\end{pmatrix}, \quad X(\lambda_2)=\begin{pmatrix}1\\-1\\1\\-1\end{pmatrix}, \quad X(\lambda_3)=\begin{pmatrix}1\\i\\-1\\-i\end{pmatrix}, \quad \text{and} \quad X(\lambda_4)=\begin{pmatrix}1\\-i\\-1\\i\end{pmatrix}.$$

(b) The characteristic polynomial is

$$P_A(t)=t^3-t^2-t-2,$$

whose unique real root is $\lambda_1=2$. The two other roots are complex, which we call λ_2 and λ_3, and they are distinct because they are conjugates. The equation $Av=\lambda v$ is equivalent to the system

$$\begin{cases} -x+z=\lambda x \\ -x+3y=\lambda y \\ -4x+13y-z=\lambda z. \end{cases}$$

When $\lambda = 2$, a basis for V_2 is $X(\lambda_1) = \begin{pmatrix} 1 \\ 1 \\ 3 \end{pmatrix}$.

When $\lambda = \lambda_2$ or $\lambda = \lambda_3$, then the above system is equivalent to

$$\begin{cases} z = (\lambda + 1)x \\ -x = (\lambda - 3)y \\ -4x + 13y = (\lambda + 1)z, \end{cases}$$

so the following eigenvectors form a basis for the associated eigenspace:

$$X(\lambda_2) = \begin{pmatrix} 1 \\ -1/(\lambda_2 - 3) \\ (\lambda_2 + 1) \end{pmatrix} \text{ and } X(\lambda_3) = \begin{pmatrix} 1 \\ -1/(\lambda_3 - 3) \\ (\lambda_3 + 1) \end{pmatrix}.$$

8. *Find the eigenvalues and a basis for the eigenspaces for the following matrices.*

(a) $\begin{pmatrix} 2 & 4 \\ 5 & 3 \end{pmatrix}$ *(b)* $\begin{pmatrix} 1 & 2 \\ 2 & -2 \end{pmatrix}$ *(c)* $\begin{pmatrix} 3 & 2 \\ -2 & 3 \end{pmatrix}$

(d) $\begin{pmatrix} -1 & 2 & 2 \\ 2 & 2 & 2 \\ -3 & -6 & -6 \end{pmatrix}$ *(e)* $\begin{pmatrix} 3 & 2 & 1 \\ 0 & 1 & 2 \\ 0 & 1 & -1 \end{pmatrix}$ *(f)* $\begin{pmatrix} -1 & 4 & -2 \\ -3 & 4 & 0 \\ -3 & 1 & 3 \end{pmatrix}$.

SOLUTION. (a) The characteristic polynomial is

$$P_A(t) = t^2 - 5t - 14,$$

so the eigenvalues are $\lambda_1 = 7$ and $\lambda_2 = -2$. Then the equation $AX = \lambda X$ is equivalent to

$$\begin{cases} 2x + 4y = \lambda x \\ 5x + 3y = \lambda y. \end{cases}$$

So if we put $x = 4$, we see that a basis for V_{λ_1} is $X(\lambda_1) = \begin{pmatrix} 4 \\ 5 \end{pmatrix}$.

Putting $x = 1$, we find that a basis for V_{λ_2} is $X(\lambda_2) = \begin{pmatrix} 1 \\ -1 \end{pmatrix}$.

(b) The characteristic polynomial is

$$P_A(t) = t^2 + t - 6,$$

so the eigenvalues are $\lambda_1 = 2$ and $\lambda_2 = -3$. Then the equation $AX = \lambda X$ is equivalent to

$$\begin{cases} x + 2y = \lambda x \\ 2x - 2y = \lambda y. \end{cases}$$

So if we put $y = 1$, we see that a basis for V_{λ_1} is $X(\lambda_1) = \begin{pmatrix} 2 \\ 1 \end{pmatrix}$.

Putting $x = 1$, we see that a basis for V_{λ_2} is $X(\lambda_2) = \begin{pmatrix} 1 \\ -2 \end{pmatrix}$.

(c) The characteristic polynomial is

$$P_A(t) = t^2 - 6t + 13,$$

so the eigenvalues are $\lambda_1 = 3 + 2i$ and $\lambda_2 = 3 - 2i$. Then the equation $AX = \lambda X$ is equivalent to

$$\begin{cases} 3x + 2y = \lambda x \\ -2x + 3y = \lambda y \end{cases} \Leftrightarrow \begin{cases} 2y = (\lambda - 3)x \\ 2x = (3 - \lambda)y. \end{cases}$$

When $x = 1$, we see that bases for V_{λ_1} and V_{λ_2}, respectively, are given by

$$X(\lambda_1) = \begin{pmatrix} 1 \\ (\lambda_1 - 3)/2 \end{pmatrix} = \begin{pmatrix} 1 \\ i \end{pmatrix} \quad \text{and} \quad X(\lambda_2) = \begin{pmatrix} 1 \\ (\lambda_2 - 3)/2 \end{pmatrix} = \begin{pmatrix} 1 \\ -i \end{pmatrix}.$$

(d) The characteristic polynomial is

$$P_A(t) = t(t^2 + 5t + 6),$$

so the eigenvalues are $\lambda_1 = 0$, $\lambda_2 = -2$, and $\lambda_3 = -3$. Then $Ax = \lambda x$ is equivalent to

$$\begin{cases} -x + 2y + 2z = \lambda x \\ 2x + 2y + 2z = \lambda y \\ -3x - 6y - 6z = \lambda z \end{cases} \Leftrightarrow \begin{cases} 2y + 2z = (\lambda + 1)x \\ 2x + 2z = (\lambda - 2)y \\ -3x - 6y = (\lambda + 6)z. \end{cases}$$

When $\lambda = 0$, the first and second equations imply that $x = 0$, so a basis for V_{λ_1} is

$$X(\lambda_1) = \begin{pmatrix} 0 \\ 1 \\ -1 \end{pmatrix}.$$

When $\lambda = -2$, the first and second equations that imply $z = 0$; so if we put $y = 1$, we see that a basis for the eigenspace V_{λ_2} is

$$X(\lambda_2) = \begin{pmatrix} -2 \\ 1 \\ 0 \end{pmatrix}.$$

When $\lambda = -3$, the first and second equations imply that $y = 0$; so if we put $x = 1$ we find that a basis for V_{λ_3} is

$$X(\lambda_3) = \begin{pmatrix} 1 \\ 0 \\ -1 \end{pmatrix}.$$

(e) The characteristic polynomial is

$$P_A(t) = t^3 - 3t^2 - 3t + 9 = (t-3)(t^2-3),$$

so the eigenvalues are $\lambda_1 = 3$, $\lambda_2 = \sqrt{3}$, and $\lambda_3 = -\sqrt{3}$. The equation $AX = \lambda X$ is equivalent to

$$\begin{cases} 3x + 2y + z = \lambda x \\ \quad y + 2z = \lambda y \\ \quad y - z = \lambda z \end{cases} \Leftrightarrow \begin{cases} 2y + z = (\lambda - 3)x \\ \quad 2z = (\lambda - 1)y \\ \quad y = (\lambda + 1)z. \end{cases}$$

So when $\lambda = 3$, we find that $y = z = 0$; so a basis for V_{λ_1} is $\begin{pmatrix} 1 \\ 0 \\ 0 \end{pmatrix}$.

When $\lambda = \lambda_2$ or $\lambda = \lambda_3$, we set $z = 1$, so that bases for V_{λ_2} and V_{λ_3} are respectively, given by

$$\begin{pmatrix} \dfrac{2(\lambda_2+1)+1}{\lambda_2-3} \\ \lambda_2+1 \\ 1 \end{pmatrix} = \begin{pmatrix} \dfrac{2+\sqrt{3}}{1-\sqrt{3}} \\ \sqrt{3}+1 \\ 1 \end{pmatrix} \quad \text{and} \quad \begin{pmatrix} \dfrac{2(\lambda_3+1)+1}{\lambda_3-3} \\ \lambda_3+1 \\ 1 \end{pmatrix} = \begin{pmatrix} \dfrac{2-\sqrt{3}}{1+\sqrt{3}} \\ -\sqrt{3}+1 \\ 1 \end{pmatrix}.$$

(f) The characteristic polynomial is

$$P_A(t) = t^3 - 6t^2 + 11t - 6 = (t-1)(t-2)(t-3),$$

so the eigenvalues are $\lambda_1 = 1$, $\lambda_2 = 2$, and $\lambda_3 = 3$. The equation $AX = \lambda X$ is equivalent to

$$\begin{cases} -x + 4y - 2z = \lambda x \\ \quad -3x + 4y = \lambda y \\ -3x + y + 3z = \lambda z. \end{cases}$$

When $\lambda = 1$, we find that $x = y$ and $z = y$; so a basis for V_{λ_1} is $\begin{pmatrix} 1 \\ 1 \\ 1 \end{pmatrix}$.

When $\lambda = 2$, we find that $z = y$ and $3x = 2y$; so a basis for V_{λ_2} is $\begin{pmatrix} 2 \\ 3 \\ 3 \end{pmatrix}$.

When $\lambda = 3$, we find that $y = 3x$ and $4x = z$; so a basis for V_{λ_3} is $\begin{pmatrix} 1 \\ 3 \\ 4 \end{pmatrix}$.

9. *Let V be an n-dimensional vector space and assume that the characteristic polynomial of a linear map $A\colon V \to V$ has n distinct roots. Show that V has a basis consisting of eigenvectors.*

SOLUTION. Each root of the characteristic polynomial is an eigenvalue which has a non-zero eigenvector. The n eigenvectors associated to the eigenvalues are linearly independent because the eigenvalues are distinct so these eigenvectors form a basis for V.

10. *Let A be a square matrix. Show that the eigenvalues of ${}^tA$ are the same as those of A.*

SOLUTION. The transpose of $\lambda I - A$ is $\lambda I - {}^tA$. Since the determinant of a matrix is equal to the determinant of its transpose, we conclude that the eigenvalues of ${}^tA$ are the same as those of A.

11. *Let A be an invertible matrix. If λ is an eigenvalue of A show that $\lambda \neq 0$ and that λ^{-1} is an eigenvalue of A^{-1}.*

SOLUTION. If $\lambda = 0$, then for some non-zero vector v we would have $Av = 0$, which is impossible because A is invertible. Therefore, $\lambda \neq 0$ and we have

$$v = A^{-1}Av = A^{-1}(\lambda v) = \lambda A^{-1}(v),$$

hence λ^{-1} is an eigenvalue of A^{-1}.

12. *Let V be the space over* **R** *generated by the two functions* $\sin t$ *and* $\cos t$. *Does the derivative (viewed as a linear map of V into itself) have any non-zero eigenvectors in V? If so, which?*

SOLUTION. The matrix of the derivative with respect to $\{\sin t, \cos t\}$ is

$$\begin{pmatrix} 0 & -1 \\ 1 & 0 \end{pmatrix},$$

so the characteristic polynomial, which is $P_D(t) = t^2 + 1$, has no real root. Therefore, the derivative does not have a non-zero eigenvector. One could also proceed directly from the definition. Let $v = a\cos t + b\sin t$. Then $Dv = \lambda v$ implies

$$\begin{cases} -a = \lambda b \\ \ \ \ b = \lambda a. \end{cases}$$

So $(\lambda^2 + 1)a = 0$, which implies that $a = b = 0$; so we see that D does not have a non-zero eigenvector in V.

13. *Let D denote the derivative which we view as a linear map on the space of differentiable functions. Let k be an integer $\neq 0$. Show that the functions* $\sin kx$ *and* $\cos kx$ *are eigenvectors for* D^2. *What are the eigenvalues?*

SOLUTION. We see that the eigenvalue is $-k^2$ because

$$D(\sin kx) = k\cos kx \text{ so } D^2(\sin kx) = -k^2 \sin kx$$

and similarly, $D^2(\cos kx) = -k^2 \cos kx$.

14. *Let* $A: V \to V$ *be a linear map of V into itself, and let* $\{v_1, \ldots, v_n\}$ *be a basis of V consisting of eigenvectors having distinct eigenvalues* $c_1, \ldots, c_n$. *Show that any eigenvector v of A in V is a scalar multiple of some* v_i.

SOLUTION. We assume $v \neq 0$. Then we can write $v = \sum_{i=1}^{n} a_i v_i$, where not all a_i are zero. Then $Av = \lambda v$ implies

$$\sum_{i=1}^{n} a_i c_i v_i = \sum_{i=1}^{n} \lambda a_i v_i;$$

so

$$\sum_{i=1}^{n} a_i (c_i - \lambda) v_i = O.$$

Since not all a_i are zero, we see that $c_k = \lambda$ for some k. Since all c_i are distinct, we must have $a_i = 0$ for all $i \neq k$, proving that v is some scalar multiple of v_k.

15. *Let A, B be square matrices of the same size. Show that the eigenvalues of AB are the same as the eigenvalues of BA.*

SOLUTION. We contend that the eigenvalues of *BA* are also the eigenvalues of *AB*. Suppose that for some non-zero vector *X* we have

$$BAX = \lambda X \qquad (*)$$

Case 1. If $\lambda = 0$, then *BA* is not invertible. Hence $\text{Det}(BA) = 0$, which implies that $\text{Det}(B) = 0$ or $\text{Det}(A) = 0$. Therefore, *AB* is not invertible; so for some non-zero vector *Y* we have $ABY = O$, proving that 0 is an eigenvalue of *AB*.

Case 2. Assume $\lambda \neq 0$. Then $(*)$ implies, $A(BAX) = A(\lambda X)$; so

$$(AB)(AX) = \lambda(AX).$$

The vector *AX* is non-zero; otherwise, from $(*)$ we see that $\lambda = 0$, which is a contradiction. So λ is an eigenvalue of *AB*, and this proves our contention.

By symmetry we conclude that the eigenvalues of *AB* are the same as the eigenvalues of *BA*.

VIII, §3 Eigenvalues and Eigenvectors of Symmetric Matrices

1. *Find the eigenvalues of the following matrices, and the maximum value of the associated quadratic form on the circle.*

(a) $\begin{pmatrix} 2 & -1 \\ -1 & 2 \end{pmatrix}$ *(b)* $\begin{pmatrix} 1 & 1 \\ 1 & 0 \end{pmatrix}$.

SOLUTION.
(a) The characteristic polynomial is

$$P_A(t) = t^2 - 4t + 3,$$

so the eigenvalues are 1 and 3. The maximum of the associated quadratic form on the unit circle is 3.

(b) The characteristic polynomial is

$$P_A(t) = t^2 - t - 1,$$

so the eigenvalues are

$$\frac{1+\sqrt{5}}{2} \quad \text{and} \quad \frac{1-\sqrt{5}}{2}.$$

The maximum of the associated quadratic form on the unit circle is the largest of these eigenvalues.

2. *Same question, except find the maximum on the unit sphere.*

(a) $\begin{pmatrix} 1 & -1 & 0 \\ -1 & 2 & -1 \\ 0 & -1 & 1 \end{pmatrix}$ *(b)* $\begin{pmatrix} 2 & -1 & 0 \\ -1 & 2 & -1 \\ 0 & -1 & 2 \end{pmatrix}$.

SOLUTION.
(a) The characteristic polynomial is

$$P_A(t) = t^3 - 4t^2 + 3t = t(t-1)(t-3),$$

so the eigenvalues are 0, 1 and 3. Hence the maximum of the associated quadratic form on the unit sphere is 3.

(b) The characteristic polynomial is

$$P_A(t) = t^3 - 6t^2 + 10t - 4 = (t-2)(t^2 - 4t + 2),$$

so the eigenvalues are

$$2,\ 2+\sqrt{2} \text{ and } 2-\sqrt{2},$$

and the maximum of the associated quadratic form on the unit sphere is $2+\sqrt{2}$.

3. *Find the maximum and minimum of the function*

$$f(x,y) = 3x^2 + 5xy - 4y^2$$

on the unit circle.

SOLUTION. The given function is the quadratic form associated with the matrix

$$\begin{pmatrix} 3 & \frac{5}{2} \\ \frac{5}{2} & -4 \end{pmatrix}.$$

The characteristic polynomial of this matrix is

$$P_A(t) = t^2 + t - \tfrac{73}{4},$$

so the eigenvalues are

$$\frac{-1+\sqrt{74}}{2} \quad \text{and} \quad \frac{-1-\sqrt{74}}{2}.$$

Thus the maximum of f on the unit circle is the largest eigenvalue, and the minimum of f on the unit circle is the smallest eigenvalue.

VIII, §4 Diagonalization of a Symmetric Linear Map

1. *Suppose that A is a diagonal $n \times n$ matrix. For any $X \in \mathbf{R}^n$, what is ${}^tXAX$ in terms of the coordinates of X and the diagonal elements of A?*

SOLUTION. If $\lambda_1, \ldots, \lambda_n$ are the diagonal elements of the matrix A, then

$$^tXAX = \lambda_1 x_1^2 + \lambda_2 x_2^2 + \ldots + \lambda_n x_n^2.$$

2. *Let*

$$A = \begin{pmatrix} \lambda_1 & 0 & \ldots & 0 \\ 0 & \lambda_2 & \ldots & 0 \\ \vdots & & & \vdots \\ 0 & 0 & \ldots & \lambda_n \end{pmatrix}$$

be a diagonal matrix with $\lambda_1 \geq 0, \ldots, \lambda_n \geq 0$. Show that there exists an $n \times n$ matrix B such that $B^2 = A$.

SOLUTION. Squaring the matrix

$$B = \begin{pmatrix} \sqrt{\lambda_1} & 0 & \ldots & 0 \\ 0 & \sqrt{\lambda_2} & & \vdots \\ \vdots & & & 0 \\ 0 & \ldots & 0 & \sqrt{\lambda_n} \end{pmatrix},$$

we find that $B^2 = A$.

3. *Let V be a finite dimensional vector space with a positive definite scalar product. Let $A: V \to V$ be a symmetric linear map. We say that A is* ***positive definite*** *if $\langle Av, v \rangle > 0$ for all $v \in V$ and $v \neq O$. Prove:*
(a) If A is positive definite, then all eigenvalues are > 0.
(b) If A is positive definite, then there exists a symmetric linear map B such that $B^2 = A$ and $BA = AB$. What are the eigenvalues of B? [Hint: Use a basis of V consisting of eigenvectors.]

SOLUTION. (a) Let λ be an eigenvalue for A and v a non-zero eigenvector for λ with $\|v\| = 1$. Then

$$\lambda = \langle \lambda v, v \rangle = \langle Av, v \rangle > 0.$$

(b) Let $\{v_1, \ldots, v_n\}$ be an orthonormal basis of V of eigenvectors of A with respective eigenvalues $\lambda_1, \ldots, \lambda_n$. Then the matrix of A with respect to this basis is

$$A_M = \begin{pmatrix} \lambda_1 & 0 & \dots & 0 \\ 0 & \lambda_2 & & \vdots \\ \vdots & & & 0 \\ 0 & \dots & 0 & \lambda_n \end{pmatrix}.$$

As in Exercise 2, consider the linear map $B: V \to V$ whose matrix with respect to $\{v_1, \dots, v_n\}$ is

$$B_M = \begin{pmatrix} \sqrt{\lambda_1} & 0 & \dots & 0 \\ 0 & \sqrt{\lambda_2} & & \vdots \\ \vdots & & & 0 \\ 0 & \dots & 0 & \sqrt{\lambda_n} \end{pmatrix}.$$

Then clearly we have $B_M^2 = A_M$ and $B_M A_M = A_M B_M$, and the eigenvalues of B are $\sqrt{\lambda_1}, \dots, \sqrt{\lambda_n}$. We contend that B is symmetric. If $v = \sum a_i v_i$ and $w = \sum b_i v_i$, then

$$\langle Bv, w \rangle = \sum_{i=1}^{n} a_i b_i \sqrt{\lambda_i} = \langle v, Bw \rangle,$$

which proves our contention and concludes the exercise.

4. *We say that A is* ***semipositive*** *definite if* $\langle Av, v \rangle \geq 0$ *for all* $v \in V$. *Prove the analogue of (a), (b) of Exercise 3 when A is only assumed semipositive. Thus the eigenvalues are all* ≥ 0, *and there exists a symmetric linear map B such that* $B^2 = A$.

SOLUTION. The proof is exactly the same as in Exercise 3, with $\geq$ signs instead of $>$.

5. *Assume that A is symmetric positive definite. Show that* A^2 *and* A^{-1} *are symmetric positive definite.*

SOLUTION. The map A is invertible because the assumptions imply that $\operatorname{Ker} A = \{O\}$. For A^2 we have

$$\langle A^2 v, w \rangle = \langle Av, Aw \rangle = \langle v, A^2 w \rangle$$

and

$$\langle A^2 v, v \rangle = \langle Av, Av \rangle > 0$$

whenever $v \neq O$; so A^2 is symmetric and positive definite.

For A^{-1} we have

$$\langle A^{-1}v, w\rangle = \langle A^{-1}v, AA^{-1}w\rangle = \langle AA^{-1}v, A^{-1}w\rangle = \langle v, A^{-1}w\rangle$$

and

$$\langle A^{-1}v, v\rangle = \langle A^{-1}v, AA^{-1}v\rangle = \langle A(A^{-1}v), A^{-1}v\rangle > 0$$

whenever $v \neq O$.

6. *Let* $A\colon \mathbf{R}^n \to \mathbf{R}^n$ *be an invertible linear map.*
(i) Show that ${}^tAA$ *is symmetric positive definite.*
(ii) By Exercise 3b, there is a symmetric positive definite B such that $B^2={}^tAA$. *Let* $U = AB^{-1}$. *Show that U is unitary.*
(iii) Show that $A = UB$.

SOLUTION. (i) The map ${}^tAA$ is symmetric because

$$\langle {}^tAAv, w\rangle = \langle Av, Aw\rangle = \langle v, {}^tAAw\rangle,$$

and if $v \neq O$, then $Av \neq O$; so

$$\langle {}^tAAv, v\rangle = \langle Av, Av\rangle > 0,$$

thus ${}^tAA$ is positive definite.

(ii) We have $U^{-1} = BA^{-1}$ and $BB={}^tAA$, so

$$BA^{-1} = B^{-1}\,{}^tAAA^{-1} = B^{-1}\,{}^tA,$$

but Exercise 3 implies $B^{-1}={}^tB^{-1}$, so $U^{-1}={}^tU$.

(iii) $UB = AB^{-1}B = A$.

7. *Let B be symmetric positive definite and also unitary. Show that* $B = I$.

SOLUTION. We have $B={}^tB$ and ${}^tB = B^{-1}$, so $B = B^{-1}$. Considering a basis of eigenvectors of B, and looking at the matrices of B and its inverse with respect to this basis, we find that

$$\lambda_i = \frac{1}{\lambda_i},$$

where $\lambda_1, \ldots, \lambda_n$ are the eigenvalues of the basis vectors. Since $\lambda_i > 0$, we see that $\lambda_i = 1$; so $B = I$.

8. *Prove that a symmetric real matrix A is positive definite if and only if there exists a non-singular matrix N such that $A={}^tNN$. [Hint: Use Theorem 4.4, and write ${}^tUAU$ as the square of a diagonal matrix, say B^2. Let $N = UB^{-1}$.]*

SOLUTION. Once direction was proved in Exercise 6 (i). Suppose that A is symmetric positive definite. Then ${}^tUAU$ is positive definite because U is invertible and

$${}^tX{}^tUAUX={}^t(UX)A(UX) > 0.$$

So there exists an invertible diagonal matrix B such that $B^2={}^tUAU$. Let $N = B{}^tU$; then

$${}^tNN = U{}^tBB{}^tU = UB^2{}^tU = A.$$

9. *Find an orthogonal basis of $\mathbf{R}^2$ consisting of eigenvectors of the given matrix.*

(a) $\begin{pmatrix} 1 & 3 \\ 3 & 2 \end{pmatrix}$ *(b)* $\begin{pmatrix} -1 & 1 \\ 1 & 2 \end{pmatrix}$ *(c)* $\begin{pmatrix} 2 & 0 \\ 0 & 2 \end{pmatrix}$

(d) $\begin{pmatrix} 1 & 1 \\ 1 & 1 \end{pmatrix}$ *(e)* $\begin{pmatrix} 1 & -1 \\ -1 & 1 \end{pmatrix}$ *(f)* $\begin{pmatrix} 2 & -3 \\ -3 & 1 \end{pmatrix}$.

SOLUTION. (a) The characteristic polynomial is

$$P_A(t) = (t-1)(t-2) - 9 = t^2 - 3t - 7,$$

so the eigenvalues are $\lambda_1 = \left(3+\sqrt{37}\right)/2$ and $\lambda_2 = \left(3-\sqrt{37}\right)/2$. The equation $AX = \lambda X$ is equivalent to

$$\begin{cases} 3y = (\lambda - 1)x \\ 3x = (\lambda - 2)y, \end{cases}$$

so an orthogonal basis of eigenvectors for $\mathbf{R}^2$ is given by

$$\begin{pmatrix} 1 \\ (\lambda_1 - 1)/3 \end{pmatrix} \quad \text{and} \quad \begin{pmatrix} (\lambda_2 - 1)/3 \\ 1 \end{pmatrix}.$$

(b) The characteristic polynomial is

$$P_A(t) = t^2 - t - 3,$$

so the eigenvalues are $\lambda_1 = (1+\sqrt{13})/2$ and $\lambda_2 = (1-\sqrt{13})/2$. The equation $AX = \lambda X$ is equivalent to

$$\begin{cases} y = (\lambda+1)x \\ x = (\lambda-2)y, \end{cases}$$

so an orthogonal basis of eigenvectors for $\mathbf{R}^2$ is given by the two vectors

$$\begin{pmatrix} 1 \\ \lambda_1 + 1 \end{pmatrix} \quad \text{and} \quad \begin{pmatrix} \lambda_2 - 2 \\ 1 \end{pmatrix}.$$

(c) The characteristic polynomial is

$$P_A(t) = (t-2)^2,$$

so the only eigenvalue is 2. Therefore, an orthogonal basis of eigenvectors for $\mathbf{R}^2$ is given by

$$\begin{pmatrix} 1 \\ 0 \end{pmatrix} \quad \text{and} \quad \begin{pmatrix} 0 \\ 1 \end{pmatrix}.$$

(d) The characteristic polynomial is

$$P_A(t) = t(t-2),$$

so the eigenvalues are 0 and 2. The equation $AX = \lambda X$ is equivalent to

$$\begin{cases} y = (\lambda-1)x \\ x = (\lambda-1)y, \end{cases}$$

so an orthogonal basis of eigenvector for $\mathbf{R}^2$ is given by the two vectors

$$\begin{pmatrix} 1 \\ -1 \end{pmatrix} \quad \text{and} \quad \begin{pmatrix} 1 \\ 1 \end{pmatrix}.$$

(e) The characteristic polynomial is

$$P_A(t) = t(t-2),$$

so the eigenvalues are 0 and 2. The equation $AX = \lambda X$ is equivalent to

$$\begin{cases} -y = (\lambda - 1)x \\ -x = (\lambda - 1)y, \end{cases}$$

so an orthogonal basis of eigenvector for $\mathbf{R}^2$ is given by the two vectors

$$\begin{pmatrix} 1 \\ 1 \end{pmatrix} \quad \text{and} \quad \begin{pmatrix} -1 \\ 1 \end{pmatrix}.$$

(f) The characteristic polynomial is

$$P_A(t) = t^2 - 3t - 7,$$

so the eigenvalues are $\lambda_1 = \left(3 + \sqrt{37}\right)/2$ and $\lambda_2 = \left(3 - \sqrt{37}\right)/2$. The equation $AX = \lambda X$ is equivalent to

$$\begin{cases} -3y = (\lambda - 2)x \\ -3x = (\lambda - 1)y, \end{cases}$$

so an orthogonal basis of eigenvectors for $\mathbf{R}^2$ is given by

$$\begin{pmatrix} 3 \\ 2 - \lambda_1 \end{pmatrix} \quad \text{and} \quad \begin{pmatrix} 1 - \lambda_2 \\ 3 \end{pmatrix}.$$

10. *Let A be a symmetric 2×2 real matrix. Show that if the eigenvalues of A are distinct, then their eigenvectors form an orthogonal basis of* $\mathbf{R}^2$.

SOLUTION. The eigenvectors of the distinct eigenvalues are linearly independent and therefore form a basis for $\mathbf{R}^2$. By Exercise 14, §2, and the Spectral Theorem we conclude at once that this basis is orthogonal.

11. *Let V be as in §4 (i.e. a vector space of dimension n over* $\mathbf{R}$, *with a positive definite scalar product). Let $A: V \to V$ be a symmetric linear map. Let v_1, v_2 be eigenvectors of A with eigenvalues λ_1, λ_2 respectively. If $\lambda_1 \neq \lambda_2$, show that v_1 is perpendicular to v_2.*

SOLUTION. We have

$$\lambda_1 \langle v_1, v_2 \rangle = \langle Av_1, v_2 \rangle = \langle v_1, Av_2 \rangle = \lambda_2 \langle v_1, v_2 \rangle,$$

so $\langle v_1, v_2 \rangle = 0$.

12. *Let V be as in §4 (i.e. a vector space of dimension n over* **R**, *with a positive definite scalar product). Let A: $V \to V$ be a symmetric linear map. If A has only one eigenvalue, show that every orthogonal basis of V consists of eigenvectors of A.*

SOLUTION. By the Spectral Theorem, there exists an orthonormal basis $\{v_1, \ldots, v_n\}$ of eigenvectors of A. Then, given any vector w, we can write $w = \sum a_i v_i$. Hence

$$Aw = \sum A(a_i v_i) = \lambda \sum a_i v_i = \lambda w,$$

where λ is the unique eigenvalue of A. Conclude.

13. *Let V be as in §4 (i.e. a vector space of dimension n over* **R**, *with a positive definite scalar product). Let A: $V \to V$ be a symmetric linear map. Let* $\dim V = n$ *and assume that there are n distinct eigenvalues of A. Show that their eigenvectors form an orthogonal basis of V.*

SOLUTION. The eigenvectors are linearly independent, so we must show that v_i is perpendicular to v_j whenever $i \neq j$. This result was proved in Exercise 11.

14. *Let V be as in §4 (i.e. a vector space of dimension n over* **R**, *with a positive definite scalar product). Let A: $V \to V$ be a symmetric linear map. If the kernel of A is $\{O\}$, then no eigenvalue of A is equal to 0, and conversely.*

SOLUTION. If $\operatorname{Ker} A = \{O\}$, there exists no non-zero vector v such that $Av = O$; hence 0 is not an eigenvalue. Conversely, if the kernel of A were not O, then A would have a non-zero vector v such that $Av = O$; hence A has 0 as an eigenvalue.

15. *Let V be as in §4 (i.e. a vector space of dimension n over* **R**, *with a positive definite scalar product). Let A: $V \to V$ be a symmetric linear map. Prove that the following conditions on A imply each other.*
(a) All eigenvalues are > 0.
(b) For all elements $v \in V$, $v \neq O$, *we have* $\langle Av, v \rangle > 0$.

If the map A satisfies these conditions, it is said to be ***positive definite***. *Thus the second condition, in terms of coordinate vectors and the ordinary scalar product in* $\mathbf{R}^n$ *reads:*

For all vectors $X \in \mathbf{R}^n$, $X \neq O$, *we have* ${}^tXAX > 0$.

SOLUTION. (b) implies (a) was proved in Exercise 3. We now prove that (a) implies (b). We can find an orthonormal basis $\{v_1, \ldots, v_n\}$ of eigenvectors of A. So if we write $v = \sum a_i v_i$, then

$$\langle Av, v\rangle = \left\langle \sum a_i \lambda_i v_i, \sum a_i v_i \right\rangle = \sum_{i=1}^{n} \lambda_i a_i^2 .$$

So $\langle Av, v\rangle > 0$ whenever $v \neq O$.

16. *Determine which of the following matrices are positive definite.*

(a) $\begin{pmatrix} 1 & 2 \\ 2 & 1 \end{pmatrix}$ *(b)* $\begin{pmatrix} 1 & -1 \\ -1 & 2 \end{pmatrix}$ *(c)* $\begin{pmatrix} 3 & 2 \\ 2 & 1 \end{pmatrix}$

(d) $\begin{pmatrix} 1 & 2 & 3 \\ 2 & 0 & 1 \\ 3 & 1 & 1 \end{pmatrix}$ *(e)* $\begin{pmatrix} 1 & -1 & 0 \\ -1 & 0 & 1 \\ 0 & 1 & 2 \end{pmatrix}$.

SOLUTION. The matrices of (c) and (e) are positive definite. The matrices of (a), (b), and (d) are not positive definite.

17. *Prove that the following conditions concerning a real symmetric matrix are equivalent. A matrix satisfying these conditions is called* ***negative definite.***

(a) All eigenvalues of A are < 0.

(b) For all vectors $X \in \mathbf{R}^n$, $X \neq O$, *we have* ${}^tXAX < 0$.

SOLUTION. Consider the linear map associated to the matrix and the ordinary inner product. Then the argument runs as in Exercise 15.

18. *Let A be an* $n \times n$ *non-singular real symmetric matrix. Prove the following statements*

(a) If λ *is an eigenvalue of A, then* $\lambda \neq 0$.

(b) If λ *is an eigenvalue of A, then* λ^{-1} *is an eigenvalue of* A^{-1}.

(c) The matrices A and A^{-1} *have the same set of eigenvectors.*

SOLUTION.

(a) Since A is invertible, $\operatorname{Ker} A = \{O\}$; so $\lambda \neq 0$.

(b) If $Av = \lambda v$, then $A^{-1}Av = \lambda A^{-1}v$; so $\lambda^{-1}v = A^{-1}v$.

(c) In (b) we see that an eigenvector for A is an eigenvector for A^{-1}. Conversely, if $A^{-1}v = kv$, then $k^{-1}v = Av$; so an eigenvector for A^{-1} is also an eigenvector for A.

19. *Let A be a symmetric positive definite real matrix. Show that A^{-1} exists and is positive definite.*

SOLUTION. See Exercise 5.

20. *Let V be as in §4 (i.e. a vector space of dimension n over* **R**, *with a positive definite scalar product). Let A and B be two symmetric operators of V such that $AB = BA$. Show that there exists an orthogonal basis of V which consists of eigenvectors for both A and B. [Hint: If λ is an eigenvalue of A, and V_λ consists of all $v \in V$ such that $Av = \lambda v$. Show that BV_λ is contained in V_λ. This reduces the problem to the case when $A = \lambda I$.]*

SOLUTION. If $v \in V_\lambda$, then we have $BAv = \lambda Bv$; so $ABv = \lambda Bv$, and thus $Bv \in V_\lambda$. So V_λ has an orthogonal basis consisting of eigenvectors of B. These vectors are also eigenvectors for A.

Now choose a basis $\{v_1, \dots, v_n\}$ for V consisting of eigenvectors of A, and let $\lambda_1, \dots, \lambda_k$ be the distinct eigenvalues of these basis vectors (cf. Exercise 21). Then we can find an orthogonal basis for V_{λ_i} that consists of eigenvectors for both A and B. Taking the union of all the bases for $V_{\lambda_1}, \dots, V_{\lambda_k}$, we see that we get an orthogonal basis for V which consists of eigenvectors of A and B.

21. *Let V be as in §4 (i.e. a vector space of dimension n over* **R**, *with a positive definite scalar product), and let $A: V \to V$ be a symmetric operator. Let $\lambda_1, \dots, \lambda_r$ be the distinct eigenvalues of A. If λ is an eigenvalue of A, let $V_\lambda(A)$ consist of the set of all $v \in V$ such that $Av = \lambda v$.*
(a) Show that $V_\lambda(A)$ is a subspace of V, and that A maps $V_\lambda(A)$ into itself. We call $V_\lambda(A)$ the ***eigenspace*** *of A belonging to λ.*
(b) Show that V is the direct sum of the space $V = V_{\lambda_1}(A) \oplus \cdots \oplus V_{\lambda_r}(A)$.
(c) Let λ_1, λ_2 be two distinct eigenvalues. Show that V_{λ_1} is orthogonal to V_{λ_2}.

SOLUTION.
(a) Clearly, $O \in V_\lambda(A)$. If $v_1, v_2 \in V_\lambda(A)$, then

$$A(v_1 + v_2) = Av_1 + Av_2 = \lambda(v_1 + v_2),$$

and if c is a scalar,

$$A(cv) = cAv = \lambda cv,$$

so $V_\lambda(A)$ is a subspace of V. If $v \in V_\lambda(A)$, then we have

$$AAv = A(\lambda v) = \lambda Av,$$

whence A maps $V_\lambda(A)$ into itself.

(b) By the Spectral Theorem, V has a orthonormal basis of eigenvectors of A, so $V = V_{\lambda_1} + \ldots + V_{\lambda_r}$. Suppose that $v = v_1 + \ldots + v_r$ and $v = w_1 + \ldots + w_r$, where $v_i, w_i \in V_{\lambda_i}$. If $u_i = v_i - w_i$, then

$$O = u_1 + \ldots + u_r.$$

But the u_i's have distinct eigenvalues, so we must have $u_i = O$ for all i.

(c) See Exercise 11.

22. *If P_1, P_2 are two symmetric positive definite real matrices (of the same size) and t, u are positive real numbers, show that $tP_1 + uP_2$ is symmetric positive definite.*

SOLUTION. The matrix $tP_1 + uP_2$ is symmetric because

$${}^t(tP_1 + uP_2) = t\,{}^tP_1 + u\,{}^tP_2 = tP_1 + uP_2.$$

The sum $tP_1 + uP_2$ is positive definite because

$${}^tX(tP_1 + uP_2)X = t\,{}^tXP_1X + u\,{}^tXP_2X > 0$$

whenever $X \neq O$.

23. *Let V be as in §4 (i.e. a vector space of dimension n over* **R**, *with a positive definite scalar product), and let $A\colon V \to V$ be a symmetric operator. Let $\lambda_1, \ldots, \lambda_r$ be the distinct eigenvalues of A. Show that*

$$(A - \lambda_1 I) \cdots (A - \lambda_r I) = O.$$

SOLUTION. Let $\{v_1, \ldots, v_n\}$ be an orthonormal basis of V consisting of eigenvectors of A. The map $L = (A - \lambda_1 I) \cdots (A - \lambda_r I)$ is linear, and for all i we have

$$Lv_i = O,$$

because one of the factors is zero.

24. *Let V be as in §4 (i.e. a vector space of dimension n over* **R**, *with a positive definite scalar product), and let A: V → V be a symmetric operator. A subspace W of V is said to be* ***invariant*** *or* ***stable*** *under A if* $Aw \in W$ *for all* $w \in W$, *i.e.* $AW \subset W$. *Prove that if A has no invariant subspaces other than O and V, then* $A = \lambda I$ *for some number* λ. *[Hint: Show first that A has only one eigenvalue.]*

SOLUTION. Suppose that A has two distinct eigenvalues λ_1 and λ_2. Then $\dim V_{\lambda_1} \geq 1$, $\dim V_{\lambda_2} \geq 1$, and both V_{λ_1} and V_{λ_2} are stable under A, so we must have $V_{\lambda_1} = V_{\lambda_2} = V$. But $V_{\lambda_1} \cap V_{\lambda_2} = \{O\}$, so we get a contradiction. Therefore, A has a unique eigenvalue, say λ. By the Spectral Theorem, V has a basis $\{v_1, \ldots, v_n\}$ of eigenvectors of A. If $v = \sum a_i v_i$, then

$$Av = \sum a_i A v_i = \lambda \sum a_i v_i = \lambda v,$$

whence $A = \lambda I$ as was to be shown.

25. *(For those who have read Sylvester's theorem.) Let A: V → V be a symmetric linear map. Referring back to Sylvester's theorem, show that the index of nullity of the form* $(v, w) \mapsto \langle Av, w \rangle$ *is equal to the dimension of the kernel of A. Show that the index of positivity is equal to the number of eigenvectors in a spectral basis having a positive eigenvalue.*

SOLUTION. Using the notation of Theorem 8.1 of Chapter V, we contend that $V_0 = \operatorname{Ker} A$. Suppose $v \in V_0$; then for all $w \in V$ we have

$$0 = (v, w) = \langle Av, w \rangle,$$

so $Av = O$. Clearly, if $v \in \operatorname{Ker} A$, then $(v, w) = 0$ for all $w \in V$. So the index of nullity of the form (,) is equal to the dimension of the kernel of A.

Let $\{v_1, \ldots, v_n\}$ be a spectral basis. Then

$$(v_i, v_i) = \langle Av_i, v_i \rangle = \lambda_i \langle v_i, v_i \rangle;$$

so $(v_i, v_i) > 0$ if and only if $\lambda_i > 0$. Conclude.

VIII, §5 The Hermitian Case

Throughout these exercises, we assume that V is a finite dimensional vector space over $\mathbf{C}$, with a positive definite scalar product. Also we assume $\dim V > 0$.
Let $A: V \to V$ be a hermitian operator. We define A to be ***positive definite*** *if*

$$\langle Av, v\rangle > 0 \quad \textit{for all } v \in V,\ v \neq O.$$

Also, we define A to be ***semipositive*** *or* ***semidefinite*** *if*

$$\langle Av, v\rangle \geq 0 \quad \textit{for all } v \in V.$$

1. *(a) If A is positive definite then all eigenvalues are > 0.*
(b) If A is positive definite, then there exists a hermitian linear map B such that $B^2 = A$ and $BA = AB$. What are the eigenvalues of B? [Hint: See Exercise 3 of §4.]

SOLUTION. (a) Let λ be an eigenvalue of A and v a corresponding non-zero eigenvector. Then

$$0 < \langle Av, v\rangle = \lambda\langle v, v\rangle,$$

so $\lambda > 0$.

(b) Choose a spectral basis $\{v_1, \ldots, v_n\}$ of eigenvectors of A, and let B be the hermitian linear map defined by the matrix

$$B_M = \begin{pmatrix} \sqrt{\lambda_1} & 0 & \ldots & 0 \\ 0 & \sqrt{\lambda_2} & & \vdots \\ \vdots & & & 0 \\ 0 & \ldots & 0 & \sqrt{\lambda_n} \end{pmatrix}$$

with respect to the basis. Then $B^2 = A$ and $AB = BA$. The eigenvalues of B are the square roots of the eigenvalues of A.

2. *Prove the analogues of (a) and (b) in Exercise 1 when A is only assumed to be semidefinite.*

SOLUTION. Replace the sign $>$ by $\geq$ in the solution of Exercise 1.

3. *Assume that A is hermitian positive definite. Show that A^2 and A^{-1} are hermitian positive definite.*

SOLUTION. The kernel of A is $\{O\}$, so A is invertible. The rest of the proof is the same as the solution to Exercise 5 of §4.

4. *Let $A: V \to V$ be an arbitrary invertible operator. Show that there exists a complex unitary operator U and a hermitian positive definite operator P such that $A = UP$. [Hint: Let P be a hermitian positive definite operator such that $P^2 = A^*A$. Let $U = AP^{-1}$. Show that U is unitary.]*

SOLUTION. Proceed as in Exercise 6 of §4. The map A^*A is hermitian and positive definite. So by Exercise 1 (b) there exists a hermitian positive definite operator P such that $P^2 = A^*A$. Let $U = AP^{-1}$. Then U is unitary because

$$PP = A^*A \quad \Rightarrow \quad PA^{-1} = P^{-1}A^*,$$

and $P^{-1}A^* = \left(P^{-1}\right)^* A^*$; so we see that $U^{-1} = U^*$. Clearly, $A = UP$.

5. *Let A be a non-singular complex matrix. Show that A is hermitian positive definite if and only if there exists a non-singular matrix N such that $A = N^*N$.*

SOLUTION. Replace ${}^tN$ by N^* in Exercise 8 of §4.

6. *Show that the matrix*

$$A = \begin{pmatrix} 1 & i \\ -i & 1 \end{pmatrix}$$

is semipositive, and find a square root.

SOLUTION. The characteristic polynomial is

$$P_A(t) = t(t-2),$$

so the eigenvalues are 0 and 2. Hence the matrix is semipositive. The equation $AX = \lambda X$ is equivalent to

$$\begin{cases} x + iy = \lambda x \\ -ix + y = \lambda y, \end{cases}$$

so an eigenvector for 0 is $w_1 = \begin{pmatrix} -i \\ 1 \end{pmatrix}$ and an eigenvector for 2 is $w_2 = \begin{pmatrix} i \\ 1 \end{pmatrix}$.

Let $U = \begin{pmatrix} -i & i \\ 1 & 1 \end{pmatrix}$, then $U^{-1} = \dfrac{1}{2}\begin{pmatrix} i & 1 \\ -i & 1 \end{pmatrix}$ and

$$U^{-1}AU = \begin{pmatrix} 0 & 0 \\ 0 & 2 \end{pmatrix};$$

so

$$U\begin{pmatrix} 0 & 0 \\ 0 & \sqrt{2} \end{pmatrix}U^{-1} = \frac{\sqrt{2}}{2}\begin{pmatrix} 1 & i \\ -i & 1 \end{pmatrix}$$

is a square root for A. Note that A is not unitary because the basis was not orthonormal.

7. *Find a unitary matrix U such that $U^{*}AU$ is diagonal, when A is equal to:*

(a) $\begin{pmatrix} 2 & 1+i \\ 1-i & 1 \end{pmatrix}$ *(b)* $\begin{pmatrix} 1 & i \\ -i & 1 \end{pmatrix}$

SOLUTION. The characteristic polynomial is

$$P_A(t) = t(t-3),$$

so the eigenvalues are 0 and 3. The equation $AX = \lambda X$ is equivalent to

$$\begin{cases} 2x + (1+i)y = \lambda x \\ (1-i)x + y = \lambda y, \end{cases}$$

so an orthogonal basis is given by the two vectors

$$w_1 = \begin{pmatrix} 1 \\ i-1 \end{pmatrix} \quad \text{and} \quad w_2 = \begin{pmatrix} 1+i \\ 1 \end{pmatrix}.$$

Dividing by the norm of each vector, we see that an orthonormal basis is given by the vectors

$$v_1 = \begin{pmatrix} 1/\sqrt{3} \\ (i-1)/\sqrt{3} \end{pmatrix} \quad \text{and} \quad v_2 = \begin{pmatrix} (1+i)/\sqrt{3} \\ 1/\sqrt{3} \end{pmatrix}.$$

So $U = \begin{pmatrix} 1/\sqrt{3} & (1+i)/\sqrt{3} \\ (i-1)/\sqrt{3} & 1/\sqrt{3} \end{pmatrix}$ solves the problem.

(b) The characteristic polynomial is

$$P_A(t) = t(t-2),$$

so the eigenvalues are 0 and 2. The equation $AX = \lambda X$ is equivalent to

$$\begin{cases} x + iy = \lambda x \\ -ix + y = \lambda y, \end{cases}$$

so orthogonal eigenvectors are given by $w_1 = \begin{pmatrix} -i \\ 1 \end{pmatrix}$ and $w_2 = \begin{pmatrix} i \\ 1 \end{pmatrix}$; an orthonormal basis of eigenvectors is given by

$$v_1 = \begin{pmatrix} -i/\sqrt{2} \\ 1/\sqrt{2} \end{pmatrix} \quad \text{and} \quad v_2 = \begin{pmatrix} i/\sqrt{2} \\ 1/\sqrt{2} \end{pmatrix}.$$

The matrix $U = \begin{pmatrix} -i/\sqrt{2} & i/\sqrt{2} \\ 1/\sqrt{2} & 1/\sqrt{2} \end{pmatrix}$ solves the problem.

8. *Let* $A\colon V \to V$ *be a hermitian operator. Show that there exists semipositive operators* P_1, P_2 *such that* $A = P_1 - P_2$.

SOLUTION. Select a spectral basis for A and let its matrix representation with respect to this basis be

$$A_M = \begin{pmatrix} \lambda_1 & 0 & \dots & 0 \\ 0 & \lambda_2 & & \vdots \\ \vdots & & & 0 \\ 0 & \dots & 0 & \lambda_n \end{pmatrix}.$$

After a reordering we may assume that $\lambda_1, \dots, \lambda_r$ are positive, $\lambda_{r+1}, \dots, \lambda_s$ are negative, and that $\lambda_{s+1}, \dots, \lambda_n$ are 0. Then consider the linear maps P_1 and P_2 whose matrices, with respect to the spectral basis are

$$P_1 = \begin{pmatrix} \lambda_1 & 0 & \dots & 0 \\ 0 & \lambda_r & & \vdots \\ \vdots & & & 0 \\ 0 & \dots & 0 & 0 \end{pmatrix} \quad \text{and} \quad P_2 = \begin{pmatrix} 0 & 0 & \dots & 0 \\ 0 & -\lambda_{r+1} & & \vdots \\ \vdots & & -\lambda_s & 0 \\ 0 & \dots & 0 & 0 \end{pmatrix},$$

respectively. Then both operators are semipositive and $A = P_1 - P_2$.

9. *An operator $A: V \to V$ is said to be normal if $AA^* = A^*A$.*
(a) Let A, B be normal operators such that $AB = BA$. Show that AB is normal.
(b) If A is normal state and prove a spectral theorem for A. [Hint for the proof: Find a common eigenvector for A and A^*.]

SOLUTION. (a) By part (b) and Exercise 20, §4, we see that we can find a basis for V consisting of eigenvectors of A, B, A^*, and B^*. The matrix representations of these four operators with respect to the chosen basis are diagonal, so we see at once that AB is normal.

(b) **Spectral Theorem for Normal Operators.** Let $A: V \to V$ be a normal operator. Then V has an orthogonal basis consisting of eigenvectors of A.

Proof. Let λ be a eigenvalue of A with non-zero eigenvector v. We contend that v is an eigenvector for A^* with eigenvalue $\bar{\lambda}$. Expanding the expression $\langle A^*v - \bar{\lambda}v, A^*v - \bar{\lambda}v\rangle$, we find

$$\langle A^*v, A^*v\rangle - \langle A^*v, \bar{\lambda}v\rangle - \langle \bar{\lambda}v, A^*v\rangle + \langle \bar{\lambda}v, \bar{\lambda}v\rangle. \quad (*)$$

The first term of $(*)$ is equal to

$$\langle A^*v, A^*v\rangle = \langle AA^*v, v\rangle = \overline{\langle v, A^*Av\rangle} = \langle Av, Av\rangle = |\lambda|^2\langle v, v\rangle.$$

The second term of $(*)$ is equal to

$$\langle A^*v, \bar{\lambda}v\rangle = \overline{\langle \bar{\lambda}v, A^*v\rangle} = \overline{\langle \bar{\lambda}Av, v\rangle} = |\lambda|^2\langle v, v\rangle.$$

The third term of $(*)$ is also equal to $|\lambda|^2\langle v, v\rangle$, so we see that

$$\langle A^*v - \bar{\lambda}v, A^*v - \bar{\lambda}v\rangle = 0,$$

which proves our contention.

Let E_λ be the space generated by v. In order to proceed as in the proof of the Spectral Theorem for symmetric operators, we must show that both

E_λ and $E_\lambda^\perp$ are A-invariant and A^*-invariant. Clearly, E_λ is A-invariant and A^*-invariant because v is an eigenvector for both A and A^*. We prove

$E_\lambda^\perp$ is A-invariant. Let $w \in E_\lambda^\perp$ and $u \in E_\lambda$. Then

$$\langle Aw, u\rangle = \langle w, A^*u\rangle = \lambda\langle w, u\rangle = 0,$$

so $A(E_\lambda^\perp) \subset E_\lambda^\perp$.

$E_\lambda^\perp$ is A^*-invariant. Let $w \in E_\lambda^\perp$ and $u \in E_\lambda$. Then

$$\langle A^*w, u\rangle = \overline{\langle Au, w\rangle} = \bar{\lambda}\langle w, u\rangle = 0,$$

whence $A^*(E_\lambda^\perp) \subset E_\lambda^\perp$. The proof then proceeds as in the Spectral Theorem.

10. *Show that the complex matrix*

$$\begin{pmatrix} i & -i \\ -i & i \end{pmatrix}$$

is normal, but is not hermitian and is not unitary.

SOLUTION. If A is the given matrix, then

$$A^* = \begin{pmatrix} -i & i \\ i & -i \end{pmatrix},$$

so we verify at once that

$$AA^* = A^*A = \begin{pmatrix} 2 & -2 \\ -2 & 2 \end{pmatrix}.$$

The matrix A is not hermitian because $A \neq A^*$, and A is not unitary because $A^*A \neq I$.

CHAPTER IX

Polynomials and Matrices

IX, §2 Polynomials of Matrices and Linear Maps

1. *Compute* $f(A)$ *when* $f(t) = t^3 - 2t + 1$ *and* $A = \begin{pmatrix} -1 & 1 \\ 2 & 4 \end{pmatrix}$.

SOLUTION. $f(A) = \begin{pmatrix} 6 & 13 \\ 26 & 71 \end{pmatrix}$.

2. *Let A be a symmetric matrix, and let f be a polynomial with real coefficients. Show that* $f(A)$ *is symmetric.*

SOLUTION. We work out the complex case in Exercise 3. In the real case, delete the complex conjugate bars.

3. *Let A be a hermitian matrix, and let f be a polynomial with real coefficients. Show that* $f(A)$ *is hermitian.*

SOLUTION. We use induction to prove that the powers of a hermitian matrix are hermitian. Clearly the assertion is true when $n = 1$. If A^{n-1} is hermitian, then

$$A^n = AA^{n-1} = {}^t\bar{A}\,{}^t\overline{A^{n-1}} = {}^t\overline{A^n}.$$

Since the sum of hermitian matrices is hermitian, and the identity matrix is hermitian, we see that $f(A)$ is hermitian.

4. *Let A, B be* $n \times n$ *matrices in a field K, and assume that B is invertible. Show that*

$$(B^{-1}AB)^n = B^{-1}A^nB$$

for all positive integers n.

SOLUTION. Induction. The statement is true when $n=1$. Suppose the statement is true for an integer n; then we have

$$\left(B^{-1}AB\right)^{n+1}=\left(B^{-1}AB\right)\left(B^{-1}AB\right)^{n}=\left(B^{-1}AB\right)\left(B^{-1}A^{n}B\right)=B^{-1}A^{n+1}B,$$

which ends the proof.

5. *Let* $f\in K[t]$. *Let A, B be as in Exercise 4. Show that*

$$f\left(B^{-1}AB\right)=B^{-1}f(A)B.$$

SOLUTION. Write $f(t)=a_n t^n+\ldots+a_0$, where $a_n\neq 0$. Then Exercise 4 implies

$$f\left(B^{-1}AB\right)=a_n\left(B^{-1}AB\right)^n+\ldots+a_0 I=a_n B^{-1}A^n B+\ldots+a_0 I.$$

But $I=B^{-1}IB$; hence

$$f\left(B^{-1}AB\right)=B^{-1}\left(a_n A^n+\ldots+a_0 I\right)B=B^{-1}f(A)B.$$

CHAPTER X

Triangulation of Matrices and Linear Maps

X, §1 Existence of Triangulation

1. *Let A be an upper triangular matrix:*

$$A = \begin{pmatrix} a_{11} & a_{12} & \dots & a_{1n} \\ 0 & a_{22} & \dots & a_{2n} \\ \vdots & \vdots & & \vdots \\ 0 & 0 & \dots & a_{nn} \end{pmatrix}.$$

Viewing A as a linear map, what are the eigenvalues of A^2, A^3 in general A^r where r is an integer ≥ 1?

SOLUTION. Induction shows that A^r is of the form

$$A^r = \begin{pmatrix} a_{11}^r & & & \\ 0 & a_{22}^r & & \\ \vdots & & & \\ 0 & \dots & 0 & a_{nn}^r \end{pmatrix}.$$

Expanding according to the first column the determinant of $tI - A^r$ we see that the characteristic polynomial of A^r is

$$P_{A^r}(t) = (t - a_{11}^r)\cdots(t - a_{nn}^r),$$

so the eigenvalues of A^r are $a_{11}^r, \dots, a_{nn}^r$.

2. *Let A be a square matrix. We say that A is* **nilpotent** *if there exists an integer $r \geq 1$ such that $A^r = O$. Show that is A is nilpotent, then all eigenvalues of A are 0.*

SOLUTION. Let λ be an eigenvalue of A, and let v be a non-zero eigenvector corresponding to λ. Then we have $Av = \lambda v$, and by induction $A^n v = \lambda^n v$ for all positive integer n. So

$$O = A^r v = \lambda^r v,$$

thus $\lambda = 0$. Note that one could also use Corollary 1.4, Exercise 4, §2 of Chapter IX, and Exercise 1 to get another proof.

3. *Let V be a finite dimensional space over the complex numbers, and let $A\colon V \to V$ be a linear map. Assume that all the eigenvalues of A are equal to 0. Show that A is nilpotent.*

SOLUTION. Select a basis such that the matrix of A with respect to this basis is upper triangular, namely,

$$A_M = \begin{pmatrix} a_{11} & a_{12} & \cdots & a_{1n} \\ 0 & a_{22} & & a_{2n} \\ \vdots & & & \vdots \\ 0 & \cdots & 0 & a_{nn} \end{pmatrix}.$$

The numbers $a_{11}, \ldots, a_{nn}$ are eigenvalues of A, so $a_{11} = \ldots = a_{nn} = 0$; thus A_M is strictly upper triangular. See Exercise 35, §3 of Chapter II, and conclude.

4. *Using fans, give a proof that the inverse of an invertible triangular matrix is also triangular. In fact, if V is a finite dimensional vector space, if $A\colon V \to V$ is a linear map that is invertible, and if $\{V_1, \ldots, V_n\}$ is a fan for A, show that it is also a fan for A^{-1}.*

SOLUTION. It is sufficient to show that V_i is A^{-1}-invariant. We contend that $A(V_i) = V_i$. Consider the restriction of A, $A\colon V_i \to V_i$. Then, since $\operatorname{Ker} A = \{O\}$, Theorem 3.3 of Chapter III implies that $A(V_i) = V_i$. Therefore $A^{-1}(V_i) = V_i$, which proves that V_i is A^{-1}-invariant.

5. *Let A be a square matrix of complex numbers such that $A^r = I$ for some positive integer r. If α is an eigenvalue of A show that $\alpha^r = 1$.*

SOLUTION. The argument of Exercise 2 implies that if α is an eigenvalue and v a corresponding non-zero eigenvector, then

$$v = Iv = A^r v = \alpha^r v;$$

so $\alpha^r = 1$.

6. *Find a fan basis for the linear maps of* $\mathbf{C}^2$ *represented by the matrices*

(a) $\begin{pmatrix} 1 & 1 \\ 1 & 1 \end{pmatrix}$ *(b)* $\begin{pmatrix} 1 & i \\ 1 & i \end{pmatrix}$ *(c)* $\begin{pmatrix} 1 & 2 \\ i & i \end{pmatrix}$

SOLUTION. (a) Clearly, the vector $\begin{pmatrix} 1 \\ -1 \end{pmatrix}$ is an eigenvector corresponding to the eigenvalue 0. Let V_1 be the space generated by this vector. Then $\{V_1, \mathbf{C}^2\}$ is a fan for the given linear map.

(b) By inspection we see that $\begin{pmatrix} i \\ -1 \end{pmatrix}$ is an eigenvector corresponding to the eigenvalue 0. Then, if V_1 is the space generated by this vector, the set $\{V_1, \mathbf{C}^2\}$ is a fan for the linear map.

(c) The characteristic polynomial is

$$P(t) = t^2 - (1+i)t - i,$$

so the eigenvalues are

$$\lambda_1 = \frac{(1+i)(1+\sqrt{3})}{2} \quad \text{and} \quad \lambda_2 = \frac{(1+i)(1-\sqrt{3})}{2}.$$

The equation $AX = \lambda X$ is equivalent to

$$\begin{cases} 2y = (\lambda - 1)x \\ ix = (\lambda - i)y, \end{cases}$$

so a non-zero eigenvector is $v_1 = \begin{pmatrix} 2 \\ \lambda_1 - 1 \end{pmatrix}$. If V_1 the space generated by this vector, then $\{V_1, \mathbf{C}^2\}$ is a fan for the linear map.

7. *Prove that an operator* $A\colon V \to V$ *on a finite dimensional vector space over* $\mathbf{C}$ *can be written as a sum* $A = D + N$, *where* D *is diagonalizable and* N *is nilpotent.*

SOLUTION. Select a fan for A and then select a fan basis $\mathcal{B}$. The matrix of A with respect to $\mathcal{B}$ is upper triangular:

$$A_M = \begin{pmatrix} a_{11} & a_{12} & \dots & a_{1n} \\ 0 & a_{22} & & a_{2n} \\ \vdots & & & \vdots \\ 0 & \dots & 0 & a_{nn} \end{pmatrix}.$$

Let

$$D_M = \begin{pmatrix} a_{11} & 0 & \dots & 0 \\ 0 & a_{22} & & \vdots \\ \vdots & & & 0 \\ 0 & \dots & 0 & a_{nn} \end{pmatrix} \quad \text{and} \quad N_M = \begin{pmatrix} 0 & a_{12} & \dots & a_{1n} \\ 0 & 0 & & \vdots \\ \vdots & & & a_{n-1n} \\ 0 & \dots & 0 & 0 \end{pmatrix},$$

and let D and N be the linear maps whose respective matrices with respect to $\mathcal{B}$ are D_M and N_M. See Exercise 35, §3 of Chapter II, and conclude.

X, §3 Diagonalization of Unitary Maps

1. *Let A be a complex unitary matrix. Show that each eigenvalue of A can be written $e^{i\theta}$ with some real θ.*

SOLUTION. Consider the linear map associated with the matrix and let $\langle\, ,\, \rangle$ be the standard hermitian product. Then, if λ is an eigenvalue and v a non-zero eigenvector, we have

$$\langle v, v\rangle = \langle Av, Av\rangle = \lambda\bar{\lambda}\langle v, v\rangle = |\lambda|\langle v, v\rangle,$$

because A is unitary. Thus $|\lambda| = 1$. Another proof consists of using Corollary 3.2 and Exercise 9, §3 of Chapter VII.

2. *Let A be a complex unitary matrix. Show that there exists a diagonal matrix B and a complex unitary matrix U such that $A = U^{-1}BU$.*

SOLUTION. By Corollary 3.2 there exists a unitary matrix $\tilde{U}$ such that $B = \tilde{U}^{-1}A\tilde{U}$ is diagonal. Let $U = \tilde{U}^{-1}$; then U is unitary and $A = U^{-1}BU$.

CHAPTER XI

Polynomials and Primary Decomposition

XI, §1 The Euclidean Algorithm

1. *In each of the following cases, write* $f = qg + r$ *with* $\deg r < \deg g$.

(a) $f(t) = t^2 - 2t + 1, \quad g(t) = t - 1$

(b) $f(t) = t^3 + t - 1, \quad g(t) = t^2 + 1$

(c) $f(t) = t^3 + t, \quad g(t) = t$

(d) $f(t) = t^3 - 1, \quad g(t) = t - 1$

SOLUTION.

(a) $f(t) = (t-1)g(t)$.

(b) $f(t) = tg(t) - 1$.

(c) $f(t) = (t^2 + 1)g(t)$.

(d) $f(t) = (t^2 + t + 1)g(t)$.

2. *If* $f(t)$ *has integer coefficients, and if* $g(t)$ *has integer coefficients and leading coefficient 1, show that when we express* $f = qg + r$ *with* $\deg r < \deg g$, *the polynomials q and r also have integer coefficients.*

SOLUTION. Use induction and proceed as in Theorem 1.1. By assumption, $b_m = 1$, so $a_n b_m^{-1}$ is an integer; thus $a_n b_m^{-1} t^{n-m} + q_1$ and r have integer coefficients.

3. *Using the intermediate value theorem of calculus, show that every polynomial of odd degree over the real numbers has a root in the real numbers.*

SOLUTION. Suppose that $p(t) = a_n t' + \ldots + a_0$, where $a_n \neq 0$ and n is odd. We may assume that $a_n > 0$ because α is a root for p if and only if α is a root for $-p$. Then for $t \neq 0$ we have

$$p(t) = a_n t^n \left(1 + \frac{a_{n-1}}{a_n t} + \ldots + \frac{a_0}{a_n t^n}\right),$$

and so $\lim_{t \to \infty} p(t) = \infty$ and $\lim_{t \to -\infty} p(t) = -\infty$. The intermediate value theorem implies that p has a real root.

4. *Let $f(t) = t^n + \ldots + a_0$ be a polynomial with complex coefficients, of degree n, and let α be a root. Show that $|\alpha| \le n \cdot \max_i |a_i|$. [Hint: Write $-\alpha^n = a_{n-1}\alpha^{n-1} + \ldots + a_0$. If $|\alpha| > n \cdot \max_i |a_i|$, divide by α^n and take the absolute value, together with a simple estimate to get a contradiction.]*

SOLUTION. If $|\alpha| \le 1$, there is no problem. Assume $|\alpha| > 1$; then we can divide

$$-\alpha^n = a_{n-1}\alpha^{n-1} + \ldots + a_0$$

by α^n so that

$$-1 = a_{n-1}\alpha^{-1} + \ldots + a_0\alpha^{-n}.$$

Taking absolute values and using the triangle inequality and our assumption that $|\alpha| > 1$, we get

$$1 \le |a_{n-1}||\alpha|^{-1} + \ldots + |a_0||\alpha|^{-n} \le |a_{n-1}||\alpha|^{-1} + \ldots + |a_0||\alpha|^{-1}.$$

If $|\alpha| > n \cdot \max_i |a_i|$, then

$$1 < \frac{1}{n} + \ldots + \frac{1}{n} = 1,$$

which is a contradiction.

XI, §2 Greatest Common Divisor

1. *Show $t^n - 1$ that is divisible by $t - 1$.*

SOLUTION. $t^n - 1 = (1 + \ldots + t^{n-1})(t - 1)$.

2. *Show that $t^4 + 4$ can be factored as a product of polynomials of degree 2 with integer coefficients.*

SOLUTION. $t^4+4=(t^2+2t+2)(t^2-2t+2)$.

3. *If n is odd, find the quotient of* t^n+1 *by* $t+1$.

SOLUTION. $t^n+1=(t^{n-1}-t^{n-2}+\ldots-t+1)(t+1)$.

4. *Let A be an* $n\times n$ *matrix over the field K, and let J be the set of all polynomials* $f(t)$ *in* $K[t]$ *such that* $f(A)=O$. *Show that J is an ideal.*

SOLUTION. If f_0 is the zero polynomial, then $f_0(A)=0I=O$; so $f_0\in J$. If $f,g\in J$, then

$$(f+g)(A)=f(A)+g(A)=O,$$

so $f+g\in J$. Finally, if $\in K[t]$, then

$$(gf)(A)=g(A)f(A)=O,$$

so $gf\in J$, thereby proving that J is an ideal.

XI, §3 Unique Factorization

1. *Let f be a polynomial of degree 2 over a field K. Show that either f is irreducible over K, or f has a factorization into linear factors over K.*

SOLUTION. If f is not irreducible, then we can write

$$f=gh,$$

where $0<\deg g<2$ and $0<\deg h<2$, so $\deg h=\deg g=1$, which proves the assertion.

2. *Let f be a polynomial of degree 3 over a field K. If f is not irreducible over K, show that f has a root in K.*

SOLUTION. If f is not irreducible, we can write $f=gh$, where $0<\deg g<3$ and $0<\deg h<3$. Since $\deg g+\deg h=\deg f=3$, we see that g or h has degree 1, and therefore f has a root in K.

3. *Let* $f(t)$ *be an irreducible polynomial with leading coefficient 1 over the real numbers. Assume* $\deg f = 2$. *Show that* $f(t)$ *can be written in the form*

$$f(t) = (t-a)^2 + b^2$$

with some $a, b \in \mathbf{R}$ *and* $b \neq 0$. *Conversely, prove that any such polynomial is irreducible over* $\mathbf{R}$.

SOLUTION. Write $f(t) = t^2 - 2at + d$. If f had a root, then f would not be irreducible, so f has no root and therefore

$$(2a)^2 - 4d^2 = 4(a^2 - d^2) < 0.$$

Completing the square, we can write

$$f(t) = (t-a)^2 - a^2 + d^2.$$

Let $b = \sqrt{d^2 - a^2}$, and conclude.

Conversely, assume that $f(t) = (t-a)^2 + b^2$ and $b \neq 0$. If f were irreducible, then f could be written as a product of linear factors, and therefore f would have a root. But since $f(t) = 0$ implies

$$(t-a)^2 = -b^2,$$

we get a contradiction.

4. *Let* f *be a polynomial with complex coefficients, say*

$$f(t) = \alpha_n t^n + \ldots + \alpha_0.$$

Define its complex conjugate

$$\bar{f}(t) = \bar{\alpha}_n t^n + \ldots + \bar{\alpha}_0$$

by taking the complex conjugate of each coefficient. Show that if f, g *are in* $\mathbf{C}[t]$, *then*

$$\overline{(f+g)} = \bar{f} + \bar{g}, \quad \overline{(fg)} = \bar{f}\bar{g},$$

and if $\beta \in \mathbf{C}$, *then* $\overline{(\beta f)} = \bar{\beta}\bar{f}$.

SOLUTION. We may assume that $\deg f \geq \deg g$, so that we can write

$$g(t) = \beta_n t^n + \ldots + \beta_0 .$$

Then we have

$$\begin{aligned}\overline{(f+g)}(t) &= \overline{(\alpha_n + \beta_n)}t^n + \ldots + \overline{\alpha_0 + \beta_0} = \left(\bar{\alpha}_n + \bar{\beta}_n\right)t^n + \ldots + \bar{\alpha}_0 + \bar{\beta}_0 \\ &= \bar{f}(t) + \bar{g}(t).\end{aligned}$$

If $g(t) = \beta_m t^m + \ldots + \beta_0$, where we assumed $m \leq n$, then

$$\bar{f}(t) \cdot \bar{g}(t) = (\bar{\alpha}_n t^n + \ldots + \bar{\alpha}_0)\left(\bar{\beta}_m t^m + \ldots + \bar{\beta}_0\right) = c_{n+m} t^{n+m} + \ldots + c_0 ,$$

where $c_k = \sum_{i=0}^{k} \bar{\alpha}_i \bar{\beta}_{k-i}$. But $\bar{\alpha}_i \bar{\beta}_{k-i} = \overline{\alpha_i \beta_{k-i}}$, so $\bar{f}(t) \cdot \bar{g}(t) = \overline{(fg)}(t)$. Finally, if $\beta \in \mathbf{C}$, then

$$\overline{(\beta f)}(t) = \overline{(\beta \alpha_n)}t^n + \ldots + \overline{\beta \alpha_0} = \bar{\beta}\bar{\alpha}_n t^n + \ldots + \bar{\beta}\bar{\alpha}_0 = \bar{\beta}\bar{f}(t).$$

5. *Let $f(t)$ be a polynomial with real coefficients. Let α be a root of f, which is complex but not real. Show that $\bar{\alpha}$ is also a root of f.*

SOLUTION. Write $f(t) = a_n t^n + \ldots + a_0$. Then $f(\alpha) = 0$ implies

$$0 = a_n \alpha^n + \ldots + a_0 .$$

Taking the complex conjugate of the above expression and noting that $\bar{0} = 0$, we get

$$0 = \bar{a}_n \bar{\alpha}^n + \ldots + \bar{a}_0 = a_n \bar{\alpha}^n + \ldots + a_0 ;$$

so $f(\bar{\alpha}) = 0$.

6. *Terminology being as in Exercise 5, show that the multiplicity of α in f is the same as that of $\bar{\alpha}$.*

SOLUTION. Let m be the multiplicity of α in f, and let $p(t) = (t - \alpha)$. Then we can write

$$f = p^m g ,$$

where $g(\alpha) \neq 0$. Hence

$$f = \bar{f} = \bar{p}^m \bar{g}.$$

Suppose that $\bar{g}(\bar{\alpha}) = 0$. Then $g(\alpha) = 0$, which is a contradiction, so the multiplicity of $\bar{\alpha}$ in f is also m.

7. *Let A be an $n \times n$ matrix in a field K. Let J be the set of polynomials f in $K[t]$ such that $f(A) = O$. Show that J is an ideal. The monic generator of J is called the* ***minimal*** *polynomial of A over K. A similar definition is made if A is a linear map of a finite dimensional vector space into itself.*

SOLUTION. See Exercise 4, §2.

8. *Let V be a finite dimensional vector space over K. Let $A: V \to V$ be a linear map. Let f be its minimal polynomial. If A can be diagonalized (i.e. if there exists a basis of V consisting of eigenvectors of A), show that the minimal polynomial is equal to the product*

$$(t - \alpha_1) \cdots (t - \alpha_r),$$

where $\alpha_1, \ldots, \alpha_r$ are the distinct eigenvalues of A.

SOLUTION. Let $\mu(t) = (t - \alpha_1) \cdots (t - \alpha_r)$. Then $\mu(A) = O$, so $\mu \in J$, where J is the ideal of all polynomials of $K[t]$ such that $f(A) = O$. We assert that μ is a generator for J. Let $f \in J$; then we can write $A = BDB^{-1}$, where D is diagonal and the diagonal elements of D are eigenvalues of A. If $f(t) = \sum a_n t^n$, then

$$O = f(A) = \sum a_n \left(BDB^{-1}\right)^n = B\left(\sum a_n D^n\right)B^{-1}.$$

So $\sum a_n D^n = O$, hence $\alpha_1, \ldots, \alpha_r$ are roots of f. This implies that μ is the monic generator for J.

9. *Show that the following polynomials have no multiple roots in* **C**.
(a) $t^4 + t$ *(b)* $t^5 - 5t + 1$
(c) any polynomial $t^2 + bt + c$ *if b, c are numbers such that* $b^2 - 4c$ *is not 0.*

SOLUTION. (a) Let $p(t) = t^4 + t$, and suppose that α is a root of p. Then

$$\alpha\left(\alpha^3 + 1\right) = 0.$$

But $p'(t)=4t^3+1$. Therefore if $\alpha=0$, then $p'(\alpha)=1$; and if $\alpha^3=-1$, then $p'(\alpha)=-3$, so α has multiplicity 1.

(b) Let $p(t)=t^5-5t+1$. Then $p'(t)=5(t^4-1)$. Assume that α is a root of p and that $p'(\alpha)=0$. So $\alpha^4=1$ and thus

$$p(\alpha)=t-5t+1=1-4t.$$

Since $p(\alpha)=0$, we get a contradiction. So α has multiplicity 1.

(c) Let $p(t)=t^2+bt+c$. The theory of quadratic equations shows that if $b^2-4c\neq 0$, then p has two distinct roots. The same result can be proved with the technique used in (a) and (b). The only root of p' is $-b/2$, which is not a root for p because

$$p(-b/2)=\frac{-b^2+4c}{4}.$$

10. *Show that the polynomial t^n-1 has no multiple roots in* **C**. *Can you determine all the roots and give its factorization into factors of degree 1?*

SOLUTION. Since $(t^n-1)'=nt^{n-1}$, we see at once that the roots of t^n-1 are not roots of its derivative, so the roots of t^n-1 have multiplicity 1. If $z=re^{i\theta}$, then

$$z^n-1=0 \quad\Rightarrow\quad r^n e^{in\theta}=e^{0i}.$$

So we must have $r=1$ and $n\theta=2\pi k$ for some integer k. Thus the n roots are

$$\left\{1,\ e^{i\frac{2\pi}{n}},\dots,e^{i\frac{2\pi(n-1)}{n}}\right\}=\left\{e^{i\frac{2\pi k}{n}}\right\}_{0\le k\le n-1}.$$

If we write $\alpha_k=e^{i\frac{2\pi k}{n}}$, then

$$t^n-1=(t-\alpha_0)\cdots(t-\alpha_{n-1})=\prod_{k=0}^{n-1}(t-\alpha_k).$$

11. *Let f, g be polynomials in $K[t]$, and assume that they are relatively prime. Show that one can find polynomials f_1, g_1 such that*

$$\begin{vmatrix} f & g \\ f_1 & g_1 \end{vmatrix}$$

is equal to 1.

SOLUTION. The polynomial 1 is the greatest common divisor of f and g, so 1 belongs to the ideal generated by f and g. Then there exist polynomials f_1 and g_1 such that

$$fg_1 - f_1 g = 1.$$

12. *Let f_1, f_2, f_3 be polynomials in $K[t]$ and assume that they generate the unit ideal. Show that one can find polynomials f_{ij} in $K[t]$ such that the determinant*

$$\begin{vmatrix} f_1 & f_2 & f_3 \\ f_{21} & f_{22} & f_{23} \\ f_{31} & f_{32} & f_{33} \end{vmatrix}$$

is equal to 1.

SOLUTION. Let d be the greatest common divisor of f_1 and f_2. Then we can find polynomials g_1 and g_2 such that $g_1 f_1 + g_2 f_2 = d$. The greatest common divisor of d and f_3 is 1, so there exist polynomials g_3 and g_4 such that $g_3 d + g_4 f_3 = 1$. Then the determinant

$$\begin{vmatrix} f_1 & f_2 & f_3 \\ -g_2 & g_1 & 0 \\ -(f_1/d)g_4 & -(f_2/d)g_4 & g_3 \end{vmatrix}$$

is equal to one.

13. *Let α be a complex number, and let J be the set of all polynomials $f(t)$ in $K[t]$ such that $f(\alpha) = 0$. Show that J is an ideal. Assume that J is not the zero ideal. Show that the monic generator of J is irreducible.*

SOLUTION. Modifying the proof of Exercise 7, we see that J is an ideal. Moreover, the monic generator g of J is of smallest degree, so if g is reducible we find that there exists a polynomial f if degree less than the one of g and such that $f(\alpha) = 0$. This is a contradiction.

14. *Let f, g be two polynomials, written in the form*

$$f = p_1^{i_1} \cdots p_r^{i_r}$$

and

$$g = p_1^{j_1} \cdots p_r^{j_r},$$

where i_ν, j_ν are integers ≥ 0, and $p_1, \ldots, p_r$ are distinct irreducible polynomials.

(a) Show that the greatest common divisor of f and g can be expressed as a product $p_1^{k_1} \cdots p_r^{k_r}$ where $k_1, \ldots, k_r$ are integers ≥ 0. Express k_ν in terms of i_ν and j_ν.
(b) Define the least common multiple of polynomials, and express the least common multiple of f and g as a product $p_1^{k_1} \cdots p_r^{k_r}$ with integers $k_\nu \geq 0$. Express k_ν in terms of i_ν and j_ν.

SOLUTION. (a) Let $k_\nu = \inf(i_\nu, j_\nu)$, and let $h = p_1^{k_1} \cdots p_r^{k_r}$. Then clearly h divides f and g; and if q divides f and g, then there exist polynomials ξ_1 and ξ_2 such that

$$p_1^{i_1} \cdots p_r^{i_r} = \xi_1 q$$
$$p_1^{j_1} \cdots p_r^{j_r} = \xi_2 q.$$

The unique factorization theorem implies that q must divide h.

(b) **Definition.** We say that h is the least common multiple of f and g if h is a multiple of f and of g. Furthermore, if p a multiple of f and g, then p is a multiple of h.

Let $k_\nu = \sup(i_\nu, j_\nu) f_1$, and let $h = p_1^{k_1} \cdots p_r^{k_r}$. Then h is a multiple of f and g. Let q be a multiple of f and g. Then there exist polynomials ξ_1 and ξ_2 such that

$$q = \xi_1 p_1^{i_1} \cdots p_r^{i_r}$$
$$q = \xi_2 p_1^{j_1} \cdots p_r^{j_r}.$$

The unique factorization theorem implies that q is a multiple of h.

15. *Give the greatest common divisor and least common multiple of the following polynomials:*
(a) $(t-2)^3(t-3)^4(t-i)$ *and* $(t-1)(t-2)(t-3)^3$

(b) $(t^2+1)(t^2-1)$ *and* $(t+i)^3(t^3-1)$.

SOLUTION. We use Exercise 14:
(a) $\text{g.c.d} = (t-2)(t-3)^3$ and $\text{l.c.m} = (t-2)^3(t-3)^4(t-i)(t-1)$.

(b) We have

$$(t^2+1)(t^2-1) = (t-i)(t+i)(t-1)(t+1)$$

and

$$(t+i)^3(t^3-1) = (t+i)^3(t-1)(t-e^{2\pi i/3})(t-e^{4\pi i/3});$$

so

$$\text{g.c.d} = (t+i)(t-1)$$

and

$$\text{l.c.m} = (t+i)^3(t-i)(t-1)(t+1)(t-e^{2\pi i/3})(t-e^{4\pi i/3}).$$

XI, §4 Application to the Decomposition of a Vector Space

1. *In Theorem 4.1 show that the image of* $f_1(A) =$ *kernel of* $f_2(A)$.

SOLUTION. Assume that $v \in \operatorname{Im} f_1(A)$. Then there exists w such that $f_1(A)w = v$. Therefore,

$$f_2(A)v = f_2(A)f_1(A)w = f_1(A)f_2(A)w = f(A)w = O,$$

and so $v \in \operatorname{Ker} f_2(A)$. Conversely, suppose that $v \in \operatorname{Ker} f_2(A)$. Then (*) of Theorem 4.1 implies that

$$v = g_1(A)f_1(A)v + g_2(A)f_2(A)v = f_1(A)g_1(A)v;$$

thus $v \in \operatorname{Im} f_1(A)$.

2. *Let $A: V \to V$ be an operator and V finite dimensional. Suppose that $A^3 = A$. Show that V is the direct sum*

$$V = V_0 \oplus V_1 \oplus V_{-1},$$

where $V_0 = \text{Ker}\, A$, V_1 is the (+1)-eigenspace of A, and V_{-1} is the (-1) eigenspace of A.

SOLUTION. Let $f(t) = t^3 - t$, $f_0(t) = t$, $f_1(t) = t - 1$, and $f_{-1}(t) = t + 1$, so that

$$f = f_0 f_1 f_{-1}.$$

We have $f(A) = O$, and f_0, f_1, and f_{-1} are relatively prime. Therefore if $V_0 = \text{Ker}\, f_0(A)$, $V_1 = \text{Ker}\, f_1(A)$, and $V_{-1} = \text{Ker}\, f_{-1}(A)$, then

$$V = V_0 \oplus V_1 \oplus V_{-1}.$$

But V_1 is the space of all v such that $Av = v$, so V_1 is the (+1)-eigenspace of A and, similarly, V_{-1} is the (-1)-eigenspace of A.

3. *Let $A: V \to V$ be an operator and V finite dimensional. Suppose that the characteristic polynomial of A has the factorization*

$$P_A(t) = (t - \alpha_1) \cdots (t - \alpha_n),$$

where $\alpha_1, \ldots, \alpha_n$ are distinct elements of the field K. Show that V has a basis consisting of eigenvectors for A.

SOLUTION. The theorem of Hamilton-Cayley guarantees that $P_A(A) = O$, so we can write

$$V = V_1 \oplus \cdots \oplus V_n = \bigoplus_{i=1}^{n} V_i,$$

where $V_i = \text{Ker}\,(A - \alpha_i)$. So V_i is the α_i-eigenspace of A. Conclude.

XI, §5 Schur's Lemma

1. *Let V be a finite dimensional vector space over the field K, and let S be the set of all linear maps of V into itself. Show that V is a simple S-space.*

SOLUTION. Let W be a proper subspace of V, and let $\{w_1,\dots,w_n\}$ be a basis for V such that $\{w_1,\dots,w_k\}$ is a basis for W. Then the map whose matrix is

$$\begin{pmatrix} 0 & \dots & & 0 \\ \vdots & & & \vdots \\ 0 & \dots & & 0 \\ 1 & 0 & \dots & 0 \end{pmatrix}$$

with respect to $\{w_1,\dots,w_n\}$ maps w_1 onto w_n. But $w_n \notin W$; conclude.

2. *Let* $V = \mathbf{R}^2$, *let S consist of the matrix* $\begin{pmatrix} 1 & a \\ 0 & 1 \end{pmatrix}$ *viewed as linear map of V into itself. Here, a is a fixed non-zero real number. Determine all S-invariant subspaces of V.*

SOLUTION. All non-proper subspaces of $\mathbf{R}^2$ have dimension one. Suppose that W has dimension one and let $\{w_1\}$ be a basis for W. Then W is S- invariant if and only if given a non-zero real number a there exists a number λ such that $T_a(w_1) = \lambda w_1$, where T_a is the linear map associated with the matrix. We can write $w_1 = xe_1 + ye_2$ so that

$$T_a(w_1) = xe_1 + aye_1 + ye_2.$$

Since $T_a(w_1) = \lambda w_1$, we must have

$$\begin{aligned} x + ay &= \lambda x \\ y &= \lambda y. \end{aligned}$$

If y is non-zero, then $\lambda = 1$; so $ay = 0$, which is impossible. Therefore we must have $y = 0$. Then W is the subspace generated by e_1, and one verifies at once that W is S-invariant.

3. *Let V be a vector space over the field K, and let* $\{v_1,\dots,v_n\}$ *be a basis of V. For each permutation* σ *of* $\{1,\dots,n\}$ *let* $A_\sigma\colon V \to V$ *be the linear map such that*

$$A_\sigma(v_i) = v_{\sigma(i)}.$$

(a) Show that for any permutation σ, τ *we have*

$$A_\sigma A_\tau = A_{\sigma\tau}$$

and $A_{\text{id}} = I$.
(b) Show that the subspace generated by $v = v_1 + \ldots + v_n$ *is an invariant subspace for the set* S_n *consisting of all* A_σ.
(c) Show that the element v *of part (b) is an eigenvector for each* A_σ. *What is the eigenvalue of* A_σ *belonging to* v?
(d) Let $n = 2$ *and let* σ *be a permutation which is not the identity. Show that* $v_1 - v_2$ *generates a 1-dimensional subspace which is invariant under* A_σ. *Show that* $v_1 - v_2$ *is an eigenvector of* A_σ. *What is the eigenvalue?*

SOLUTION. (a) For each basis vector we have

$$A_\sigma A_\tau(v_i) = A_\sigma v_{\tau(i)} = v_{\sigma\tau(i)} = A_{\sigma\tau} v_i$$

and

$$A_{\text{id}} v_i = v_i;$$

thus $A_\sigma A_\tau = A_{\sigma\tau}$ and $A_{\text{id}} = I$.

(b) Let W be the subspace generated by v and let $w \in W$. Then there exists a number α such that $w = \alpha v$, so

$$A_\sigma w = \alpha A_\sigma(v_1 + \ldots + v_n) = \alpha\left(v_{\sigma(1)} + \ldots + v_{\sigma(n)}\right).$$

But σ is a bijection of $\{1, \ldots, n\}$, so we see that $A_\sigma w = \alpha v = w$, thus proving that W is A_σ-invariant.

(c) Putting $\alpha = 1$ in (b), we see that $A_\sigma v = v$.

(d) By assumption, $v = v_1 - v_2 \neq O$ so the space W generated by v is 1-dimensional. We have

$$A_\sigma \alpha v = \alpha A_\sigma(v_1 - v_2) = -\alpha(v_1 - v_2),$$

so W is A_σ-invariant and v is an eigenvector of A_σ with eigenvalue -1.

4. *Let* V *be a vector space over the field* K, *and let* $A: V \to V$ *be an operator. Assume that* $A^r = I$ *for some integer* $r \geq 1$. *Let* $T = I + A + \ldots + A^{r-1}$. *Let* v_0 *be an element of* V. *Show that the space generated by* Tv_0 *is an invariant subspace of* A, *and that* Tv_0 *is an eigenvector of* A. *If* $Tv_0 \neq O$, *what is the eigenvalue?*

SOLUTION. Let $Tv_0 = w_0$. Then

$$A\alpha w_0 = \alpha A(I + A + \ldots + A^{r-1})v_0 = \alpha(A + \ldots + A^{r-1} + I)v_0 = \alpha w_0,$$

so the space generated by Tv_0 is A-invariant. If Tv_0 is non-zero, then it is an eigenvector of A with eigenvalue one.

5. *Let V be a vector space over the field K, and let S be a set of operators of V. Let U, W be S-invariant subspaces of V. Show that $U + W$ and $U \cap W$ are S-invariant subspaces.*

SOLUTION. If v belongs to $U + W$, then we can write $v = u + w$, where u and w lie in U and W, respectively. Then if $A \in S$, we have

$$Av = Au + Aw.$$

But Au and Aw belong to U and W, respectively, so $U + W$ is S-invariant.

If v belongs to $U \cap W$, and if $A \in S$, then Av also belongs to U and W; so $U \cap W$ is S-invariant.

XI, §6 The Jordan Normal Form

In the following exercises, we let V be a finite dimensional vector space over the complex numbers, and we let $A\colon V \to V$ be an operator.

1. *Show that A can be written in the form $A = D + N$, where D is a diagonalizable operator, N is a nilpotent operator, and $DN = ND$.*

SOLUTION. In a Jordan basis for V with respect to A split the matrix of A as a sum $A_M = D_M + N_M$, where

$$D_M = \begin{pmatrix} \begin{bmatrix} \lambda_1 & & 0 \\ & \ddots & \\ 0 & & \lambda_1 \end{bmatrix} & & 0 \\ & \ddots & \\ 0 & & \begin{bmatrix} \lambda_s & & 0 \\ & \ddots & \\ 0 & & \lambda_s \end{bmatrix} \end{pmatrix} \quad \text{and} \quad N_M = \begin{pmatrix} \begin{bmatrix} 0 & 1 & & 0 \\ & \ddots & \ddots & \\ & & \ddots & 1 \\ 0 & & & 0 \end{bmatrix} & & 0 \\ & \ddots & \\ 0 & & \begin{bmatrix} 0 & 1 & & 0 \\ & \ddots & \ddots & \\ & & \ddots & 1 \\ 0 & & & 0 \end{bmatrix} \end{pmatrix}$$

Let D and N be the maps whose matrix representations with respect to the Jordan basis are D_M and N_M, respectively. The D is diagonalizable and N is nilpotent. Moreover, $DN = ND$.

2. *Assume that V is cyclic. Show that the subspace of V generated by the eigenvectors of A is one dimensional.*

SOLUTION. Let w be an eigenvector of A. Then w has eigenvalue α and, since $\{v, (A-\alpha I)v, \dots, (A-\alpha I)^{r-1}v\}$ is a basis for V, we can write

$$w = a_0 v + \dots + a_{r-1}(A-\alpha I)^{r-1}v$$

for some $a_0, \dots, a_{r-1} \in \mathbf{C}$. Then

$$0 = (A-\alpha I)w = a_0(A-\alpha I)v + \dots + a_{r-2}(A-\alpha I)^{r-1}v,$$

which implies that $a_0 = \dots = a_{r-2} = 0$ and therefore $(A-\alpha I)^{r-1}v$ is a basis for the subspace of V generated by eigenvectors of A.

3. *Assume that V is cyclic. Let f be a polynomial. What are the eigenvalues of $f(A)$ in terms of those of A? Same question when V is not assumed cyclic.*

SOLUTION. Choose a Jordan basis for V with respect to A, and let M be the matrix of A with respect to this basis. Then the matrix of $f(A)$ with respect to the Jordan basis is $f(M)$, which is upper triangular. If λ_{ii} is the ith diagonal entry of M, then the ith diagonal entry of $f(M)$ is $f(\lambda_{ii})$. Hence the eigenvalues of $f(A)$ are $f(\lambda)$, where λ is an eigenvalue of A. Since V is the direct sum of cyclic subspaces, the answer is the same when V is not assumed cyclic.

4. *If A is nilpotent and not O, show that A is not diagonalizable.*

SOLUTION. Suppose that A is diagonalizable. Then there exists a basis of eigenvectors of A. Since A is not O, not all eigenvalues are zero. So there exists a non-zero number λ and a non-zero vector v such that $Av = \lambda v$. So

$$A^r v = \lambda^r v,$$

which shows that $A^r v$ is never zero, which is a contradiction.

5. *Let P_A be the characteristic polynomial of A, and write it as a product*

$$. \ P_A(t) = \prod_{i=1}^{r} (t - \alpha_i)^{m_i},$$

where $\alpha_1, \ldots, \alpha_r$ *are distinct. Let f be a polynomial. Express the characteristic polynomial* $P_{f(A)}$ *as a product of factors of degree 1.*

SOLUTION. Exercise 3 implies at once that $P_{f(A)}(t) = \prod_{i=1}^{r} \left(t - f(\alpha_i)\right)^{m_i}$.

CHAPTER XII

Convex Sets

XII, §4 The Krein-Milman Theorem

1. *Let A be a vector in $\mathbf{R}^n$. Let $F: \mathbf{R}^n \to \mathbf{R}^n$ be the translation*

$$F(X) = X + A.$$

Show that if S is convex in $\mathbf{R}^n$, then $F(S)$ is also convex.

SOLUTION. Let P and Q be two points of $F(S)$, and let $P_0 = P - A$ and $Q_0 = Q - A$. Then $P_0, Q_0 \in S$, so the line segment

$$tP_0 + (1-t)Q_0$$

with $0 \le t \le 1$ belongs to S. Thus $tP_0 + (1-t)Q_0 + A$ belongs to $F(S)$, but

$$tP_0 + (1-t)Q_0 + A = t(P-A) + (1-t)(Q-A) + A = tP + (1-t)Q,$$

so $F(S)$ is convex.

2. *Let c be a number > 0, and let P be a point in $\mathbf{R}^n$. Let S be the set of points X such that $\| X - P \| < c$. Show that S is convex. Similarly, show that the set of points X such that $\| X - P \| \le c$ is convex.*

SOLUTION. By Exercise 1 we see that it is sufficient to prove the assertion when $P = O$. Suppose that $\| X \| < c$ and $\| Y \| < c$. Then, if $0 \le t \le 1$, we have

$$\| tX + (1-t)Y \| \le t\| X \| + (1-t)\| Y \| < tc + (1-t)c = c,$$

so S is convex. The set of points such that $\| X \| \le c$ is the closure of S, so it is also convex.

3. *Sketch the convex closure of the following sets of points.*

(a) $(1,2)$, $(1,-1)$, $(1,3)$, $(-1,1)$

(b) $(-1,2)$, $(2,3)$, $(-1,-1)$, $(1,0)$

SOLUTION. (a)

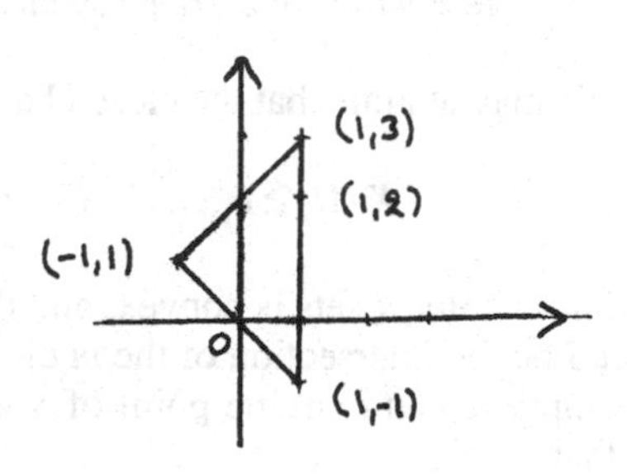

(b)

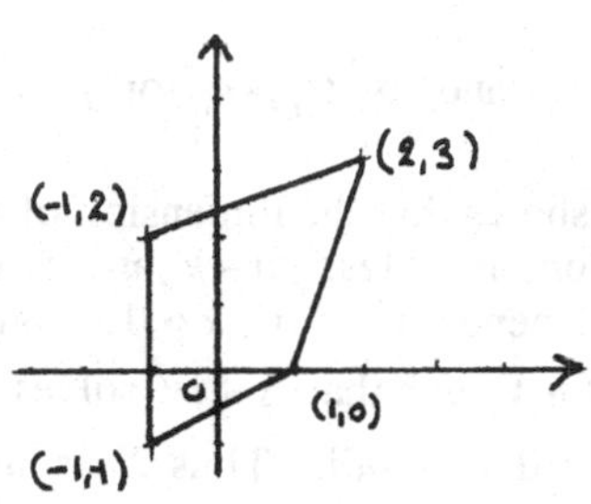

4. *Let* $L\colon \mathbf{R}^n \to \mathbf{R}^n$ *be an invertible linear map. Let S be convex in* $\mathbf{R}^n$ *and P an extreme point of S. Show that* $L(P)$ *is an extreme point of* $L(S)$*. Is the assertion still true if L is not invertible?*

SOLUTION. Suppose that $L(P)$ is not an extreme point of $L(S)$. Then there exist points Q_1, Q_2 of $L(S)$ such that $Q_1 \neq Q_2$ and such that

$$t_0 Q_1 + (1-t_0)Q_2 = L(P)$$

for some number $0 < t_0 < 1$. So

$$L^{-1}\big(t_0 Q_1 + (1-t_0)Q_2\big) = L^{-1}L(P),$$

which implies that there exist points $P_1, P_2 \in S$ with $P_1 \neq P_2$ and such that

$$t_0 P_1 + (1-t_0)P_2 = P,$$

which is a contradiction. The assertion need not hold if L is not invertible. Indeed, consider in $\mathbf{R}^2$ the map $L(x,y) = (x,0)$. Let S be the set of all

(x, y) such that $y \geq |x|$. Then S is convex and the origin is an extreme point of S, but $L(O)$ is not an extreme point of $L(S)$.

5. *Prove that the intersection of a finite number of closed half spaces in* $\mathbf{R}^n$ *can have only a finite number of extreme points.*

SOLUTION. We may assume that the closed half-spaces are described by

$$X \cdot U_1 \geq c_1, \ldots, X \cdot U_m \geq c_m.$$

The intersection of convex sets is convex, and the intersection of closed sets is closed. Let S be the intersection of the m closed half-spaces. We assume that S is nonempty, so an extreme point of S must be a boundary point of S. Suppose that

$$X \cdot U_1 = c_1, \ldots, X \cdot U_k = c_k \quad \text{and} \quad X \cdot U_j > c_j \ \text{ for } j > k \qquad (*)$$

The theory of linear equations shows that the dimension of the space solution of the first k linear equations is at least $n-k$ and is equal to $n-k$ when $U_1, \ldots, U_k$ are linearly independent. If $n-k>0$, continuity implies that there exists a non-zero vector V such that $X+tV$ solves $(*)$ for all t in some interval $[-\delta, \delta]$ where $\delta > 0$ is small. Thus X is not an extreme point. So we must have that the dimension of the space of solutions of the linear equation is 0, thus reduced to a unique point. So we see that S can have only a finite number of extreme points.

6. *Let B be a column vector in* $\mathbf{R}^n$, *and A an $n \times n$ matrix. Show that the set of solutions of the linear equations $AX = B$ is a convex set in* $\mathbf{R}^n$.

SOLUTION. We may view the matrix A as a linear map $A: \mathbf{R}^n \to \mathbf{R}^n$. Then, since B is convex, the set of X such that $AX = B$ is also convex.

Appendix

Complex Numbers

1. *Express the following complex numbers in the form $x+iy$, where x, y are real numbers.*

(a) $(-1+3i)^{-1}$ *(b)* $(1+i)(1-i)(c)$
(c) $(1+i)i(2-i)$ *(d)* $(i-1)(2-i)$
(e) $(7+\pi i)(\pi+i)$ *(f)* $(2i+1)\pi i$
(g) $(\sqrt{2}+i)(\pi+3i)$ *(h)* $(i+1)(i-2)(i+3)$.

SOLUTION.

(a) $\frac{-1}{10}-\frac{3}{10}i$ (b) 2 (c) $-1+3i$ (d) $-1+3i$

(e) $6\pi+i(7+\pi^2)$ (f) $-2\pi+i\pi$ (g) $\pi\sqrt{2}-3+(3\sqrt{2}+\pi)i$

(h) $-8-6i$.

2. *Express the following complex numbers in the form $x+iy$, where x, y are real numbers.*

(a) $(1+i)^{-1}$ *(b)* $\frac{1}{3+i}$ *(c)* $\frac{2+i}{2-i}$ *(d)* $\frac{1}{2-i}$

(e) $\frac{1+i}{i}$ *(f)* $\frac{i}{1+i}$ *(g)* $\frac{2i}{3-i}$ *(h)* $\frac{1}{-1+i}$

SOLUTION.

(a) $\frac{1}{2}-\frac{1}{2}i$ (b) $\frac{3}{10}-\frac{1}{10}i$ (c) $\frac{3}{5}+\frac{4}{5}i$ (d) $\frac{2}{5}+\frac{1}{5}i$

(e) $1-i$ (f) $\frac{1}{2}+\frac{1}{2}i$ (g) $\frac{-1}{5}+\frac{3}{5}i$ (h) $\frac{-1}{2}-\frac{1}{2}i$.

3. *Let α be a complex number $\neq 0$. What is the absolute value of $\alpha/\bar{\alpha}$? What is $\bar{\bar{\alpha}}$?*

SOLUTION. Let $\alpha = a + ib$. Then

$$\frac{\alpha}{\overline{\alpha}} = \frac{a+ib}{a-ib} = \frac{a^2 - b^2 + 2abi}{a^2 + b^2},$$

so

$$\left|\frac{\alpha}{\overline{\alpha}}\right|^2 = \frac{(a^2 - b^2)^2 + 4a^2b^2}{(a^2 + b^2)^2} = \frac{(a^2 + b^2)^2}{(a^2 + b^2)^2} = 1.$$

Moreover,

$$\overline{\overline{\alpha}} = \overline{a - ib} = a + ib = \alpha.$$

4. *Let* α, β *be two complex numbers. Show that* $\overline{\alpha\beta} = \overline{\alpha}\overline{\beta}$ *and that*

$$\overline{\alpha + \beta} = \overline{\alpha} + \overline{\beta}.$$

SOLUTION. Suppose that $\alpha = a + ib$ and $\beta = c + id$. Then

$$\overline{\alpha}\overline{\beta} = (a - ib)(c - id) = ac - bd - i(ad + bc) = \overline{\alpha\beta}.$$

and

$$\overline{\alpha} + \overline{\beta} = a - ib + c - id = a + c - i(b + d) = \overline{\alpha + \beta}.$$

5. *Show that* $|\alpha\beta| = |\alpha||\beta|$.

SOLUTION. If $\alpha = a + ib$ and $\beta = c + id$, then

$$|\alpha\beta|^2 = \alpha\beta\overline{\alpha\beta} = \alpha\beta\overline{\alpha}\overline{\beta} = \alpha\overline{\alpha}\beta\overline{\beta} = |\alpha|^2|\beta|^2.$$

6. *Define addition of n-tuples of complex numbers componentwise, and multiplication of n tuples of complex numbers by complex numbers componentwise also. If* $A = (\alpha_1, \ldots, \alpha_n)$ *and* $B = (\beta_1, \ldots, \beta_n)$ *are n-tuples of complex numbers, define their product* $\langle A, B\rangle$ *to be*

$$\alpha_1\overline{\beta}_1 + \ldots + \alpha_n\overline{\beta}_n$$

(note the complex conjugation!). Prove the following rules
HP 1. $\langle A, B\rangle = \overline{\langle B, A\rangle}$.

***HP* 2.** $\langle A, B+C\rangle = \langle A, B\rangle + \langle A, C\rangle$.
***HP* 3.** *If* α *is a complex number, then*

$$\langle \alpha A, B\rangle = \alpha\langle A, B\rangle \quad \textit{and} \quad \langle A, \alpha B\rangle = \overline{\alpha}\langle A, B\rangle.$$

***HP* 4.** *If* $A = O$ *then* $\langle A, A\rangle = 0$, *and otherwise* $\langle A, A\rangle > 0$.

SOLUTION.
HP 1. $\overline{\langle B, A\rangle} = \overline{\beta_1\overline{\alpha}_1 + \ldots + \beta_n\overline{\alpha}_n} = \overline{\beta}_1\overline{\overline{\alpha}}_1 + \ldots + \overline{\beta}_n\overline{\overline{\alpha}}_n = \langle A, B\rangle$.

HP 2. If $C = (\gamma_1, \ldots, \gamma_n)$, then

$$\begin{aligned}\langle A, B+C\rangle = \alpha_1\overline{(\beta_1 + \gamma_1)} + \ldots + \alpha_n\overline{(\beta_n + \gamma_n)} &= \alpha_1\overline{\beta}_1 + \alpha_1\overline{\gamma}_1 + \ldots + \alpha_n\overline{\beta}_n + \alpha_n\overline{\gamma}_n \\ &= \langle A, B\rangle + \langle A, C\rangle.\end{aligned}$$

HP 3. We have

$$\langle \alpha A, B\rangle = \alpha\alpha_1\overline{\beta}_1 + \ldots + \alpha\alpha_n\overline{\beta}_n = \alpha\left(\alpha_1\overline{\beta}_1 + \ldots + \alpha_n\overline{\beta}_n\right) = \alpha\langle A, B\rangle,$$

and, similarly, $\langle A, \alpha B\rangle = \overline{\alpha}\langle A, B\rangle$.

HP 4. Clearly $\langle A, A\rangle = 0$ whenever $A = 0$. Suppose that $\langle A, A\rangle = 0$. Then

$$0 = \langle A, A\rangle = |\alpha_1|^2 + \ldots + |\alpha_n|^2,$$

so $|\alpha_j| = 0$ for all j, and thus $A = 0$.

7. *We assume that you know about the functions sine and cosine, and their addition formulas. Let* θ *be a real number.*
(a) Define $e^{i\theta} = \cos\theta + i\sin\theta$. *Show that if* θ_1 *and* θ_2 *are real numbers, then*

$$e^{i(\theta_1+\theta_2)} = e^{i\theta_1}e^{i\theta_2}.$$

Show that any complex number of absolute value 1 can be written in the form e^{it} *for some real number t.*
(b) Show that any complex number can be written in the form $re^{i\theta}$ *for some real numbers* r, θ *with* $r \geq 0$.
(c) If $z_1 = r_1e^{i\theta_1}$ *and* $z_2 = r_2e^{i\theta_2}$ *with real* r_1, $r_2 \geq 0$ *and real* θ_1, θ_2, *show that*

$$z_1 z_2 = r_1 r_2 e^{i(\theta_1+\theta_2)}.$$

(d) If z is a complex number, and n an integer > 0, *show that there exists a complex number w such that* $w^n = z$. *If* $z \neq 0$ *show that there exists n distinct such complex numbers w. [Hint: If* $z = re^{i\theta}$, *consider first* $r^{1/n}e^{i\theta/n}$ *.]*

SOLUTION. (a) The addition formulas

$$\cos(\theta_1 + \theta_2) = \cos\theta_1 \cos\theta_2 - \sin\theta_1 \sin\theta_2$$
$$\sin(\theta_1 + \theta_2) = \cos\theta_1 \sin\theta_2 + \sin\theta_1 \cos\theta_2$$

imply at once that

$$e^{i\theta_1}e^{i\theta_2} = e^{i(\theta_1+\theta_2)}.$$

Let $\alpha = a + ib$ with $|\alpha| = 1$. Then we must have $a^2 + b^2 = 1$, so there exists a number t such that $a = \cos t$ and $b = \sin t$; therefore $\alpha = e^{it}$.

(b) If $z = 0$, then let $r = 0$. If $z \neq 0$, let $r = |z|$ and note that $z/|z|$ has absolute value 1.

(c) We have

$$z_1 z_2 = r_1 r_2 e^{i\theta_1} e^{i\theta_2} = r_1 r_2 e^{i(\theta_1+\theta_2)}.$$

(d) If $z = 0$, let $w = 0$. Now assume that $z \neq 0$. Since $\big| z/|z| \big| = 1$, we can write $z = re^{i\theta}$, where $r \geq 0$. We write $w = se^{i\varphi}$, where $s \geq 0$, so the equation $w^n = z$ becomes

$$s^n e^{in\varphi} = re^{i\theta}.$$

Taking absolute values we see that $s^n = r$ and $e^{i\theta} = e^{in\varphi}$ implies $n\varphi = \theta + 2k\pi$ for some integer k. Thus the n distinct roots of the equation $w^n = z$ are

$$r^{1/n}e^{i\frac{\theta}{n}},\ r^{1/n}e^{i\frac{\theta}{n}+\frac{2\pi}{n}}, \ldots, r^{1/n}e^{i\frac{\theta}{n}+\frac{2(k-1)\pi}{n}}.$$

8. *Assuming that the complex numbers are algebraically closed, prove that every irreducible polynomial over the real numbers has degree 1 or 2. [Hint: Split the polynomial over the complex numbers and pair off complex conjugate roots.]*

SOLUTION. Let P be a polynomial of $\mathbf{R}[t]$. In $\mathbf{C}[t]$, if α is a root of P, then $\overline{\alpha}$ is also a root of P. Since $\mathbf{C}$ is algebraically closed, we can factor P in $\mathbf{C}[t]$ in irreducible polynomials, namely, polynomials of degree 1; thus

$$P(t) = a_n \prod_{k=1}^{n} (t - \alpha_k),$$

where a_n is the leading coefficient of P and $\alpha_1, \ldots, \alpha_n$ are the roots in $\mathbf{C}$ of P. Let $\gamma_1, \ldots, \gamma_p$ be the real roots of P. The number of complex roots is even, so we can pair each root with its conjugate. Therefore, we let $\beta_1, \ldots, \beta_s, \overline{\beta}_1, \ldots, \overline{\beta}_s$ be the complex roots of P. Then we can write

$$P(t) = a_n \prod_{k=1}^{r} (t - \gamma_k) \prod_{j=1}^{s} (t - \beta_j)(t - \overline{\beta}_j).$$

But $(t - \beta_j)(t - \overline{\beta}_j) = t^2 - (\beta_j + \overline{\beta}_j)t + |\beta_j|^2$ and $\beta_j + \overline{\beta}_j$ is real, so we see that if a polynomial over $\mathbf{R}$ is irreducible, then it has degree 1 or 2.

For an example of an irreducible polynomial of degree 2, consider $t^2 + 1$.

9 780387 947600